SEASONS *on* MATHESON BROOK

A Journey to Self Through Exploring Nature

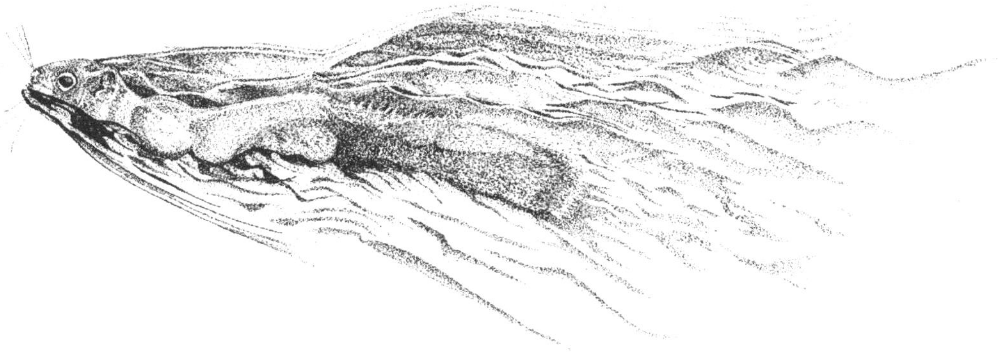

Magi Nams

Leaf Rain Books

ISBN 978-1-9990616-2-3

Copyedited by Brian Bartlett
Cover and interior design by Magi Nams
Published by Leaf Rain Books, Tatamagouche, NS
For more information, visit www.maginams.ca
Brief excerpts aired on CBC Radio One

Contents

Author's Note

In the spring of 1997, I was a harried homeschooling mother and part-time writer living in northern Nova Scotia. My training in zoology and botany seemed a lifetime ago, and I craved time to explore the natural world on my doorstep. My husband's offer to teach our sons one morning a week opened a door to freedom, and I joyfully hastened through it.

So began a year of intense observation and reflection during which I beheld natural wonders I'd never seen before and revelled in sights I saw every day. My senses opened to nature in a way they never had before, granting me a vivid appreciation for God's gift of this amazing world. Through frustration, joy, and sorrow, I scribbled notes in little notebooks while I walked, and memorized details and images while I ran. Driven to write, I drafted a rough manuscript that chronicled my explorations, reflections, and revelations. Throughout it, I sought to paint vivid images of the world around me, a world filled with intriguing and sometimes shocking natural phenomena.

In 2021, a time of anxiety and limited social interaction due to the COVID-19 pandemic, I dusted off my old journal with a view to diving deeper and with greater intimacy into how nature had sustained me during my search for self-identity twenty-four years ago. *Seasons on Matheson Brook* is my journey to self through exploring nature, and it's my love song for the earth, rural life, and family.

This book is illustrated with sixteen pointillistic pen-and-ink drawings, most completed during my year of journaling. My hope is that *Seasons on Matheson Brook*, through word and art, will encourage others to embrace the wonders of the world we live in and, in doing so, find peace and joy.

April, 2022

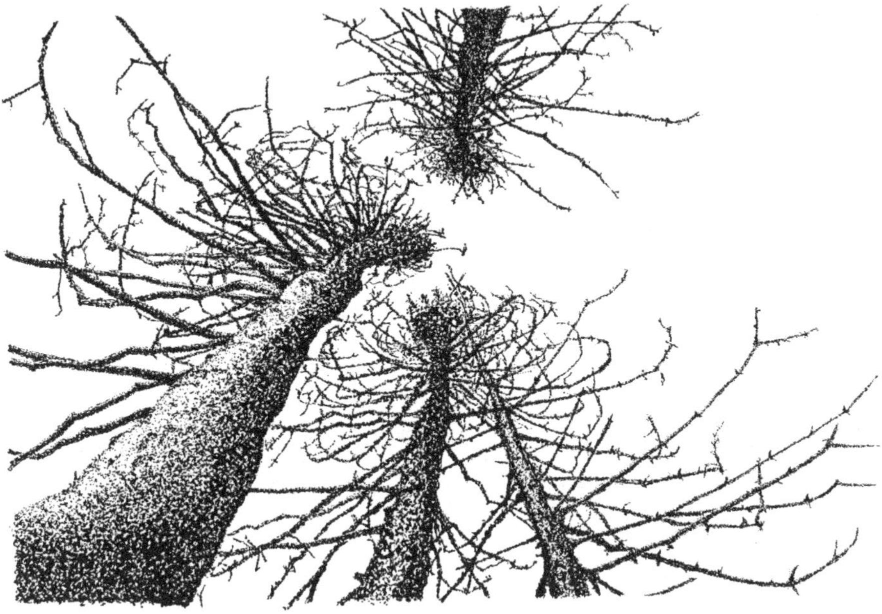

Prologue

Spring, 1997

I'm swamped. My days flash by, so cluttered with motherhood and homeschooling responsibilities that I'm not sure who I am anymore.

Each morning I rise at dawn to get an early start. I tend our nine chickens in the coop, all the while listening for the voices of songbirds returned to the forests and meadows of Ravenhill, my family's rural property in Colchester County, northern Nova Scotia. Birding is one of my passions, as is exploring other elements of the natural world. And yet, I have so little time to pursue these passions. I yearn for more.

When the chicken chores are done, I hastily don my sneakers and run or walk roads near my home, squeezing in a taste of fresh air, exercise, and solitude before my husband, Vilis, leaves for work. The remainder of my day will likely hold more fresh air and exercise, but it won't hold solitude. I drink in the sights and sounds around me as I face the day ahead.

After I return to our old farmhouse, I sweep floors and then prepare breakfast for Vilis and our young sons, Dainis and Jānis. (A note on pronunciations: Vilis rhymes with Phyllis, Magi sounds like Maggie, Dainis [DINE-is], and Jānis [YAWN-is].) Dainis and Jānis set the table, and after breakfast, feed our cats, Charcoal and Blotchy, and collect eggs from our half-dozen hens. Vilis washes the dishes, then leaves to drive to the Nova Scotia Agricultural College in Bible Hill, where he's a professor in the Environmental Sciences Department.

Soon after Vilis departs, it's time for school. My sons and I start at 7:30 a.m. with our treasured "reading time," a half-hour during which I read aloud from historical fiction, classic children's books, fun reads, or outdoor adventures. Then Dainis, who is eight, and Jānis, who is five, hit the books. They do their schoolwork in the living room or at the kitchen table or on the kitchen floor in front of the wood stove—I don't care where as long as the work gets done. We mix up their studies with stretches, calisthenics, table tennis, and walks on our property to refresh our brains and re-energize our bodies. The mornings are intense. Some run smoothly, and some are punctuated with grumblings, complaints, and tears.

We finish at noon or 1 p.m., then I update our homeschooling journal and plan the next day's assignments. Dainis and Jānis have free time for the rest of the day, although some of that time is taken up with extracurricular activities: piano lessons, homeschool gym classes, swimming lessons, skating lessons, Beavers and Cubs. The calendar on our kitchen wall is cluttered with kids' activities, and because I'm the parent at home, I'm Mom's Taxi for most of them.

I know I'm immersed in a season wherein my life focus is my children, and I know this season will pass too quickly, yet more and more my heart cries out against the total dedication required. I love my kids and, like most mothers, am proud of them, but I crave time alone, time to have my brain to myself. A freelance writer, I'm a solitary individual who is refreshed by my own company, and I need space and time to write. Those necessities are scarce in my life right now.

Even more than time to write, I long for freedom to walk beneath tall trees, identify birds, delve into Creation's wonders, and feel them spark my spirit and nourish my soul. Before I had kids, I studied biology for seven years at university, collected fossils in the foothills of the Rockies, trapped small mammals in the Northwest Territories, did botanical research in the High Arctic, helped live-trap skunks in Manitoba, and carried out vegetation surveys in the Yukon. I still want biology, or at least natural history, to be part of my life. I still want to seek what the earth would teach me, and in doing so, to discover more of myself. I enjoy sharing nature with my sons, but outings with them are not conducive to reflection.

And then there are the everyday chores that squeeze my free time into even smaller bits: housework, cooking slow food, caring for our yard, garden, and my beloved flowerbeds. All these tasks are needed. All are worthy. All are time thieves.

More days pass. I become exhausted, resentful, and so restless that I'm continually snapping at my sons and husband. Something inside me, like a song sparrow behind bars, is fighting to break free.

There is one way I could open the cage's door. We could stop homeschooling. In the midst of my frazzled existence, this seems like a ready solution. Vilis and I together chose homeschooling as our preferred educational method, but I'm the parent who does the teaching.

I ask my husband to come for a walk with me. Nearly trembling with emotion, I tell him I can't homeschool any longer. It's too much for me. I need time for myself.

Vilis listens intently, his gaze concerned. I know he's really seeing me, really hearing me. "What if I help? I can teach the boys one morning a week."

My breath catches in my throat, a glimmer of freedom appearing on my horizon. "Starting when?"

"Tomorrow."

Suddenly, I'm no longer in this alone. It's not all my responsibility anymore. The freedom I've yearned for—to write and to steep myself in the world around me and follow where it takes me—is as close as the sunrise of a new day.

1

The Earth Awakens

April 8, 1997 (4°C, cloudy, 8:30 a.m.)

Ice. Puddles. Wet brown leaves.

From the top of a steep wooded slope beside Matheson Brook, I see ice jammed upstream of our old bridge, water running brown and murky below it. Beside me, dead hemlock branches are daggers aimed at my eyes. A raven's hoarse call rips the air, and a flicker's sharp "kee-yer" slices it. Frantic squawks break out as movement swishes through a nearby thicket of young spruce trees. The action fades in a trail of diminishing vibrations. Bird? Mammal? I have no idea.

All around me, the earth awakens. Spring has come to northern Nova Scotia, and I'm searching. Who can say where we find ourselves? In the air? In the soil? In the rush of water through jumbled ice plates jammed in ridges across Matheson Brook?[i] Never during our previous five winters on our rural property, Ravenhill, did my family see ice jammed in the brook the way we saw it this past winter, and not only saw, but climbed over. Vilis or I led the way, testing the ice's thickness and strength. Dainis and Jānis followed in our careful path, stepping over gaps, the water rushing fast and black below. Now the ice-laden winter eases into spring, and the air is filled with the calls and songs of blue jays, dark-eyed juncos, black-capped chickadees, and a fox sparrow whose brilliant melody sets spring alight.

Today another brilliant melody—a melody of freedom—also sets my heart alight. For the first weekday morning since my family's academic year began eight months ago, I have no teaching responsibilities. Moments ago, Vilis sat with our boys and their schoolbooks at the kitchen table, and I stepped outdoors to explore at my leisure our seventy-eight acres of woodland and meadows.

Now I look and listen as I follow an overgrown woodcutting road that looks down onto Matheson Brook. The air is quiet except for a crow's bright call and a distant chainsaw's dull roar. On my left, tree trunks grown over rusted fence wires enclose metal barbs within bark-covered lips. On my right, a steep bank plunges to the brook. At my feet, grey, mud-stained flowers decorate wet snow. Some are eight centimetres or three inches long, and some are smaller. These are the footprints of a canid, a member of the dog family. Here coyote or fox or dog stopped to gaze out over the brook or perhaps toss a look over

---

[i] The Nova Scotia Museum refers to this waterway as Matheson Brook; it is also known as Balmoral Brook.

its shoulder. Thirteen years ago, Vilis and I cared for and released three red foxes in southwestern Manitoba. Those foxes' steps were elegant and smooth, their prints more delicate than these. My guess is that this canid was a coyote or dog, but perhaps I'm wrong.

A male ruffed grouse drums his quickening wing beats. One spring I watched a ruffie strut his breeding display, with tail fanned and black neck ruffs erected, but I've never seen one drumming. This spring I'll search again. I want to observe the initial slow wing-cups of air and to hear their measured "thump...thump...thump" increase in speed and intensity until wings and air blur sound and motion into a wild throb, as though it were the earth's heartbeat racing up a cliff and then tumbling over the edge into silence.

Ice collapses on the brook. The chainsaw's roar rubs the air with rawness. The ruffie's drumming ends abruptly and, minutes later, begins again. A flock of evening grosbeaks—the males, bright yellow and black—passes overhead. I clench my bare hands against the cold, the air temperature only a few degrees above freezing. I was in such a hurry to get out into this amazing world that I forgot to put on gloves. It's here in nature that I sense God's power and creative genius. It's here that I draw near to him. John Muir, who profoundly felt God's presence in nature, wrote, "...but when we contemplate the whole globe as one great dewdrop, striped and dotted with continents and islands, flying through space with other stars all singing and shining together as one, the whole universe appears as an infinite storm of beauty."[1]

Today I seek a place in that beautiful, infinite storm. Not for my kids and me; not for my husband and me. Only for me.

I came into this world as an individual and will exit it as an individual. In this present moment of freedom, I want to experience my individuality for all it's worth.

As I walk, my rubber boots make sucking sounds in snowy slush that trickles downslope and feeds Matheson Brook's swollen riffles. That brook is the heart of Ravenhill. Two springs ago, my family canoed it to Waughs River then paddled that river to the Northumberland Strait that separates Nova Scotia from Prince Edward Island. In summer, we walk among the brook's riffles, holding viewers made from milk cartons and clear plastic wrap to magnify caddis fly larvae in their inch-long sand-grain houses that always face upstream. We toss pebbles into the rushing water, build low rock dams at its edge, and cross it by hopping from one gleaming wet rock to another. One autumn when largetooth aspen leaves lay like golden coins on the brook, we crossed by stepping from rock to rock and began climbing the steep bank on the far side—the bank beside me now. Our male cat, Dusty, followed our footsteps precisely, but our elegant she-cat, Blotchy, mistook the leaf-covered water for solid ground and plunged into the brook before scrambling wildly to pull herself onto a rock. We laughed as she shook water distastefully from her paws before crossing safely and following us up the slope. And in winter, as in the winter past, we walk the brook's ice after Christmas when January brings consistently cold temperatures and ice freezes onto banks and stretches over the riffles, stones, and caddis fly larvae.

Here at Ravenhill, each season brings its own brook pleasures, its own enticements to explore and see what one can see. Nature holds so much beauty, so much complexity and diversity. Its webs of life intermingle in a vast and intricate balance that we, as humans, can only catch glimpses of through observation and the opening of our senses into keen awareness of our surroundings. It is this keen awareness that I wish to awaken in myself, lest one day I mistake leaf-covered water for solid ground and take an unwanted plunge into a chilling wet darkness.

The canid tracks I'm following veer to the west amid birch and spruce trees and then rejoin me as a wobbly line. I tuck my hands into my coat pockets, my fingertips stinging. Tall white spruces edge the trail on my right, and on my left, young eastern hemlocks and mature aspens reach for the sky. Needles, twigs, and lichens litter snow under the trees; this natural debris is part of an unending cycle of forest rejuvenation. Unlike human-sourced litter—such as beer cans, potato chip bags, and coffee cups strewn along roadsides—the needles, twigs, and lichens naturally return to the soil and feed a new generation of trees. As novelist Barbara Kingsolver wrote, "…the forest eats itself and lives forever."[2] And as the forest eats itself and grows anew, it feeds countless organisms and soothes the souls of humans.

The canid tracks veer into an aspen grove and disappear on ice. Farther on, I encounter a jumble of canid and snowshoe hare tracks, the latter showing imprints of large hind feet landing ahead of front feet. Was the canid hunting the hare ("rabbit" here in northern Nova Scotia), or did the two animals simply pass this way through different windows in time? I'm not skilled enough to read the story, only to identify the characters. I know that a hare's hopping motion places its hind feet ahead of where its forefeet land, but I don't know why this particular hare was in this particular place. My life isn't lived at ground level, nose to earth, with my survival dependent on the keenness of my senses. Instead, I'm an outsider, a human who claims ownership of land and water, when in truth land and water are not mine to own. They belong to the creatures of the wild who know no property boundaries, whose movements are ruled instead by an instinctive need for food, shelter, and safety. Really, I'm more caretaker than owner. This earth I walk is a gift entrusted to me. I have the privilege of spending my days on land I share with a multiplicity of wild things, including enigmatic snowshoe hares and unknown canids who leave tracks like roses imprinted in the snow.

At the base of the steep bank on my right, ice spreads a greenish mantle around the feet of apple, spruce, and hemlock trees. Farther along the trail, the ice disappears within the curved wall of what my family calls the Amphitheatre, where thick grey trunks of sugar maples, their bark scored with long narrow fissures, rise from brown leaves on the brook's flood plain.

Last summer, beavers constructed a fifty-metre dam spanning Matheson Brook, downstream from the Amphitheatre. Built of countless chewed-off aspen branches packed together with mud and rocks, the dam flooded low-lying land adjacent to the brook, potentially imperilling our maples. Where are those beavers now? And who made this wood road that winds downhill from what were once grain fields or pastures, down to the Amphitheatre and up the hill again? Our property was once a farm cleared by Scottish settlers. On misty evenings, it's easy to imagine a piper standing atop a rise, his kilt swaying as his bagpipes wail into the darkening air.

I leave the Amphitheatre and follow the wood road uphill, with patches of moss like tiny green islands in rumpled slush beneath my boots. Water glistens in a silent stream two inches wide that splits into a miniature delta near my toe. A "trickle-stream," Jānis would call this rivulet, and I delight again in his unfettered imagination one winter-spring day when he and I walked our access road and he discovered trickle-streams.

Playwright Eden Phillpotts wrote, "The universe is full of magic things, patiently waiting for our wits to grow sharper."[3] Today this forest is filled with such things: the pounding courtship song of a ruffed grouse, tall trees creaking whispered stories, meandering footprints that weave an enigma onto the trail I follow. When exploring our land with my sons, I instruct Dainis and Jānis to "look small," to see not only the long view but also what's at their fingertips or under their feet—like trickle-streams. Intriguing discoveries come in many shapes and sizes, and the more we attune our senses to the world around us at different scales of observation, the more marvels we'll see. Wonder is perhaps the single most powerful emotion to spur within us the need to preserve our natural surroundings—and also to share our findings with others so that they, too, can experience this emotion that's as fresh as a child's delighted smile in a jaded world.

A grackle creaks out its song, and a crow's wing-feathers whip the air. Deer tracks cross the wood road, their melted edges forming small, split bowls. I creep carefully up an icy bend around which my family cross-country skis in winter with excitement, trepidation, or a blend of both. Little Jānis crouches until he's almost sitting on his skis. More confident, Dainis executes a controlled snow plow, braking as he navigates the bend. I do the same, each time wondering if I'll conquer the curve without taking a tumble into the snow. Sometimes I do; sometimes I don't. Vilis never wonders. He skis the bend full throttle, with snow flying and white windsuit flapping. A glimpse into the character of any individual could be obtained simply by supplying them with cross-country skis and watching them ski this sharp curve in our old wood road.

At the top of the hill, I enter what prairie dwellers describe as parkland—trees interspersed with meadows. Our parkland is the legacy of thirty years of white spruce trees and white pine trees tossing seeds onto an old field's fertile soil, and of white-tailed deer dispersing apple seeds. More canid prints, smaller and neater than the others, mark the snow. This time I'm sure they're fox tracks following a newer trail, one hewn by my family. Winter's dregs still surround me—patches of thin snow are tongues of vanilla icing, and last year's grasses mound in pointed tussocks around stubborn goldenrod stems or curve into partly frozen dunes, creating rough, sticking-up hair on the scalp of our land.

I rush the last fifty metres to our yard, my hands and feet aching from the cold. Our rooster, Samson, crows from beneath a spruce windbreak west of the house, where his hen harem scratches and pecks at soil. Blue jays, juncos, and chickadees wing to a wooden feeder on a post in a flowerbed. I'm home.

Shivering, I sit at my desk, feeling newly alive, as though part of me long asleep is finally awake, spurred into action and vivid thought by birds, trees, and tracks in the snow. Driven to write, I begin a journal, words arising from the heart like a love song.

April 9 (-6°C, cloud/clear, 6:45 a.m.)

Frozen grass crunches under my boots. The sun, barely over the horizon, flames red onto the breast of a robin flying overhead. A woodpecker raps on a tree trunk, proclaiming its breeding territory. I hurry along the trail I walked yesterday, except in reverse, heading for the sugar maples in the Amphitheatre. Yesterday's rumpled slush is rumpled ice today, making walking precarious. I slip and nearly fall at the tricky bend in the old wood road. Somewhere out of sight ahead of me, our toboggan thumps over ice. Vilis's voice rings out in the clear air.

When I reach the Amphitheatre, Vilis and the boys are already down on the brook's flood plain, checking two-gallon aluminium sap buckets hanging on sugar maple trees. A big turquoise plastic container into which Vilis will pour maple sap is strapped to the toboggan at the top of the Amphitheatre's curved wall. I slide down the incline—not entirely of my own will—and my boots crash through ice into shallow water pooled behind a small beaver dam located upstream from the big dam that spans Matheson Brook.

Vilis approaches me, carrying a pail filled with clear liquid. Blobs of soft ice float atop the freshly gathered maple sap. "Trying to sneak up on us, were you?" he asks as he begins ascending the icy slope. He slips, like I did, and jostles the pail, causing sap to splash onto his coat and pants. "Well, that's not going to work."

Jānis's face crumples into dismay. "Now we'll have no sap!"

"How about using that?" I point to a ladder built last fall and erected against one of our maples by a deer hunter who sat high among its branches, awaiting prey. We've posted No Hunting signs on our access roads but not along our property's perimeter, with the result that hunters occasionally take advantage of the frequent use of our lower interval by white-tailed deer. Vilis and I drag the ladder to the Amphitheatre wall and brace it there. It enables him to ascend the slope with the next bucket of sap, but he slips as soon as he steps off the ladder. More sap is lost, but not all. The satisfying sound of maple essence pouring into the waiting container dances in the fresh air.

As a child growing up on a prairie farm in Alberta, far from sugar maple country, I read stories about boiling maple sap into syrup and candy. How I yearned, straight from my core to the tips of my fingers and toes, to experience that delicious harvest. Here in Nova Scotia, my childhood dream comes true every spring at Ravenhill. As daylength increases, the sun climbs higher in the sky with each passing day. Its warmth bathes the treetops of *Acer saccharum*, setting the scene for the maple sap run, which is governed purely by physical forces. When a pattern of alternating warm days and night freezes occurs, pressure within the tree stem builds up. That pressure forces sap upward and outward toward any escape route, including a sugarmaker's spile. The sap, transported in long tubes called xylem within the cambium or sapwood, contains plant sugars and nutrients that generate the distinctive sweet, delicious flavour associated with maple syrup. That much-loved flavour is forty times less concentrated in the sap than in the final product, so we must gather a large quantity if we hope to fill a pantry shelf with jars of syrup.

Again, Vilis begins climbing the ladder with a sap-filled bucket, but his wet boots slip off the rungs and once more maple sap splashes down the slope. So much for Plan B. Those future jars of syrup seem an ever-increasing distance away.

Laughing as they play, Dainis and Jānis hack-whack ice with sticks and scramble about on broad aspen trunks felled by beavers. Vilis and I stand silent, subdued by ice and defeat, but we're not done yet. Vilis scrambles up the Amphitheatre wall, ties the plastic container tightly to the toboggan, then pushes the sled down the wall. It hits bottom and, like I did, crashes through the ice, its lower end buried in slush. Vilis stands up to his rubber-booted shins in the slush and empties the next bucket of sap into the plastic tub without losing a single drop. Then he transfers a spile (also called a spigot, spout, or tap) from one maple tree to another. The mechanical whir of his cordless drill boring into the maple's sapwood is unexpected among the sounds of a woodpecker drumming, the brook rushing, and our sons' footsteps on ice. Jānis flashes me a grin when I tell him the plastic container is a quarter full—our sap haul for today.

After Vilis secures the tub's lid, he tugs the toboggan toward the Amphitheatre's wall. Dainis and Jānis cross the thin ice on a short section of beaver dam and climb the steep incline on their hands and knees, as though they were brightly coloured bears. I become a bear, too, pushing the toboggan upslope while Vilis pulls it. Then we follow the old wood road up the hill from the Amphitheatre, fighting to keep our footing where yesterday I walked so easily among moss forests and trickle-streams.

In the parkland beyond the woods, the toboggan slides smoothly over frosted grass and patches of ice. Working together, Vilis and I pull it as though we were a team of Clydesdales, Belgians, or Percherons. In my mind, I see the massive draft horses in the MacMillan Show Centre in Bible Hill, their broad chests and strong legs straining to move a thousand pounds or more of weight on a sledge in the horse pull contest at the Nova Scotia Provincial Exhibition. The horses strain. Vilis and I strain, and behind us, in the plastic container strapped to the toboggan, the sloshing gurgle of maple sap proclaims victory over gravity and ice and heralds the reward of full jars of maple syrup for our pantry.

April 10 (-6°C, overcast with light snow, W/SW breeze, 9:15 a.m.)

I'm walking what my family calls the "back road," tracing the path I ran early this morning in pursuit of cardio-pulmonary fitness. The road is a dirt and gravel right-of-way that provides all-season access to our property. From our yard, it winds through 200 metres of meadows and parkland before entering mixed-wood forest dominated by towering hemlocks. In the woods, I see where my sneakers slipped on ice—first my right foot and then my left—just as I've seen skidding tracks made by hares, squirrels, and canids who slipped on ice. Ten steps farther on, I encounter a confusion of canid tracks, mere suggestions now, dusted with new snow. What drew the canid here, to this very spot in the middle of the road? Did another canid earlier pass this way? Did a prey animal pause here?

I walk through a world of mystery, a book filled with pages written in scripts I can't read or barely recognize. All around me, animals, plants, and other organisms are living out their lives, yet I see and understand so little of their reality, as though I occupy a different existential plane. And yet, if in the next minute society were to collapse and all my civilized amenities were to disappear, my current plane of existence would crash abruptly to the forest floor. I would come face to face with raw mortality, and in that scenario my survival would depend in large part on learning to read scripts written by non-human residents of this earth.

A blue jay's split-pitch cry rings out. A pileated woodpecker barks its loud "wek-wek-wek." All around me trees creak gently, awakened to motion by wind that sighs deeply above me. How many times have I run this road? Sometimes only up and down its 800-metre length; sometimes, as the first part of a four-kilometre loop that takes me through these woods and then onto the pavement of Balmoral Road, the gravel of Matheson Brook Road, and finally, our 400-metre "front road" that parallels Matheson Brook, crosses it, then climbs a steep hill to our yard. Conquering that last hill is the final goal, the upward rise, the crowning achievement of pounding heart and straining legs. Each time I run the loop, I tell myself that if I can make it up the last hill to our yard, I can do anything.

But I'm here now, in the peace of the backroad woods where my own footprints lead me on. I slipped again there—more skid marks. Running on ice is a cautious venture, with toes spreading wide, hips bracing, and strides, short and wary. A return to gravel brings instant relaxation, with hips loosening, toes pointing forward, and legs stretching in longer strides. Do the hares, squirrels, and canids whose skidding tracks I've seen feel the same tension and release? As a child, I watched the Walt Disney movie *Bambi* on the big screen, and even now I can recall the scene where Thumper, a rabbit, entices the fawn, Bambi, onto an ice-covered pond to play. Gleefully, big-footed Thumper slides onto the ice and frolics across it, but Bambi can't keep his stilt-like legs under himself on the slippery surface. Even as a child—long before I learned zoological terms for animal locomotion—I recognized that animals possess adaptations in tune with their lifestyle, and one suite of those adaptations concerns the structure of feet.

In general, animals show three types of locomotion: plantigrade, digitigrade, and unguligrade. Like hares and squirrels, bears and raccoons, humans show plantigrade locomotion. With each step I take, my weight and balance are supported by all my foot bones, from heel to toes, pressing flat against the ground surface. This "flat-footed" mode of movement provides a large surface area to grip the ground surface, which explains why Thumper, with his large plantigrade paws, was at ease hopping and sliding on the ice-covered pond.

Now, in the movie, there was no fox pup on the ice with Thumper and Bambi, but if there had been, I might have noticed that that fox pup's feet looked different from Thumper's, but not so extremely different as Bambi's. Foxes, coyotes, and other members of the dog family exhibit digitigrade locomotion, as do members of the cat family. These animals walk with their weight balanced on their toes and the balls of their feet. Their ankles and wrists are permanently elevated off the ground and have elongated tarsal bones and carpal bones, the bones of the ankle and wrist, respectively. Although this foot morphology results in less surface area in contact with the ground surface, it also lengthens the leg, thus allowing for greater speed—useful for running down prey.

And what about Bambi, whose legs skidded out from under him? Deer, like cattle, horses, and antelopes, exhibit unguligrade locomotion, that is, they walk on the tips of their toes, which bear hoofs. The ankle and wrist bones of these animals are even more elongated than in animals with digitigrade locomotion, causing their ankles and wrists to be elevated even higher. The result is even longer legs, and the potential for even greater speed—useful for outrunning predators. The downside is that animals with unguligrade locomotion have a very small proportion of their entire foot surface in contact with the ground, which is why Bambi had such difficulty maneuvering on ice.

Beside me, as I stand on the back road, snow gathers on roadside hemlock saplings, creating frosted planes of fragile green needles on angled, graceful branches. I want to draw those branches, to find that same grace, those same angles somewhere in me. Winter is the season of branches, with leaves gone, needles subdued, and snow lining the tops of boughs horizontal and thrusting, boughs reaching upward, boughs spreading wide like the sheltering arms of a mother. In Lewis Carroll's *Alice's Adventures in Wonderland*, Alice muses to her kitten in Looking-glass House, "I wonder if the snow *loves* the trees and fields, that it kisses them so gently? And then it covers them up snug, you know, with a white quilt; and perhaps it says 'go to sleep, darlings, till the summer comes again.'"[4] In the backroad woods, I see a thousand quilts. No, *ten* thousand. The entire forest is an artwork of white on dark grey that tells stories of tree structure and comeliness I never see in summer. I long to apply pen and ink to paper to capture this striking image, to relay this winter story offered so freely here in the backroad forest.

Two ravens race past overhead in swift, air-brushed flight. One raven's voice is higher, one lower, and a third raven flies in pursuit. Can this still be courtship? I thought it would be earlier, but perhaps I'm wrong. Ravens are among the earliest nesters in Nova Scotia, typically beginning their breeding season in early March.

Even with insulated boots and gloves, I'm cold, so I strike out on the road again, walking in the shade of hemlock giants. The air is damp, as it so often is in Nova Scotia's maritime climate, yet my discomfort fades as I gaze at uplifted branches, their lacy green fingertips intertwining overhead. I've reached what my family calls the Cathedral, a grove of majestic hemlocks whose boughs arch over the road like a cathedral ceiling. The hemlocks are graceful, strong, and, if it's possible for a tree to be proud, they are proud. Through them, my spirit rises to the sky. I've stood in manmade cathedrals in Europe and felt a sense of awe. In this natural cathedral, I don't feel awe; rather, I feel peace and something more—a something more that makes me run tall, that wakes my awareness each time I pass through this forest, and that makes me ask who I am to receive a gift of such beauty.

Seneca, a Roman philosopher who lived during the time of Christ and who advised the emperor Nero for more than a decade, wrote in one of his moral epistles: "When you enter a grove peopled with ancient trees, higher than the ordinary, and shutting out the sky with their thickly inter-twined branches, don't the stately shadows of the wood, the stillness of the place, and the awful gloom of this doomed cavern then strike you with the presence of a deity?"[5] The apostle Paul, who lived concurrently with Seneca, wrote in his epistle to early Roman Christians: "By taking a long and thoughtful look at what God has created, people have always been able to see what their eyes as such cannot see: eternal power, for instance, and the mystery of his divine being..."[6] Here in the Cathedral, standing among thick trunks that are like columns holding the sky, I'm always struck by the presence of a deity, always aware of God's unfathomableness and power.

Past the Cathedral, the forest is more open, thinned by loggers. Thankfully, their chainsaws were halted by an orange flag and yellow slash at the Cathedral's edge. It's an unfortunate fact that my family doesn't own the Cathedral. The lofty hemlocks crown a property sandwiched between our seventy-eight acres to the north and the thinned forest to the south. Although the Cathedral isn't ours, we live in hope that it won't be taken from us. Occasionally its owner comes with his pickup and chainsaw to cut firewood. I see his truck, hear his saw's roar, and wonder if his awareness comes alive each time that he enters this forest, if within these woods he's struck by God's presence, and if he knows about running tall.

Back at the house, two grackles—sleek black cousins of starlings and orioles—peck for fallen sunflower seeds below the feeder in our front yard flowerbed. The grackles' yellow eyes provide a striking contrast to their iridescent purple heads and necks. I point out the birds and their iridescence to my sons.

Dainis's brow furrows. "What does iridescent mean?"

My mind claws its way back to an Ornithology class I took two decades ago while studying zoology at the University of Alberta in Edmonton. *Feather. Rachis. Barb. Barbule.* I explain that iridescent colour is produced when light is bounced around and reflected from the surfaces of interlocking barbs and barbules in a feather's vane, which lies on both sides of the feather's shaft or rachis. The purple of the grackles' head and neck feathers is seen only when light falls onto the feathers from certain angles. "That's why we see flashes of purple when the grackles move," I tell Dainis. "Now, do you see the bird in the feeder, the one with the pinky-purple head and chest?"

He nods.

"It's a purple finch. Its colours are the result of pigments in its feathers. Pigments are microscopic granules of colour. The purple finch's colour is on the *inside* of its feathers. The grackle's shiny purple colour is on the *outside* of its feathers."

Iridescent colour. Pigment colour. Understanding lights Dainis's eyes.

April 12 (late afternoon)

Maple fragrance, sweet and rich, wafts from a canning pot resting on a ring of stones that encloses a hot fire. The sap we collected this morning is on its way to becoming maple syrup. It takes forty litres of sap to produce one litre of syrup, and the rendering process involves hours of evaporation. Our sap boils over, froths down the canner's blackened sides, and drips into the fire. A pungent yet sweet scent of scorched maple fills the air. Vilis removes the pot from the fire, and Dainis and Jānis throw dried grasses and goldenrod stems onto the flames. Smoke billows up, as thick and yellow-white as sheep's wool. I sit on a picnic table near the fire pit, watching, listening, gathering images. The scent of maple is everywhere.

What is it about harvests that so completely immerses us in our surroundings—the sights, sounds, smells, and tastes? Is it a sense of urgency? Having grown up on a farm, I know there are times when undoubtedly that is true. Come harvest time on the Canadian Prairies, weather rules all. If I were on the Alberta farm of my youth on a sunny autumn day, Thomas Lake would sparkle bluer than the sky, and swaths of wheat, barley, and oats would lie in the sun like silver ribbons draped over the fields' gentle slopes. Chaff would spit from the combine's spreader, and pinkish-brown wheat kernels would dance a rattle dance in the combine's auger and then spray in heavy free fall into the grain truck's box. By evening, the dry mealy scent of grain dust would permeate the air, and sunset would be red and thick from that dust. Combine lights would reach into the falling darkness to light the way past midnight and on into deepest night until dew settled on the swaths. Then the harvesters—my father and brothers—would drive wearily home to their beds.

Three years ago, I took my sons home to the farm during harvest. Jānis was only two, and Dainis, five. They sat in their

Uncle Rick's lap to steer the combine and rode in the grain truck with their Uncle Hank. They chewed wheat gum, climbed in and out of the swather, and watched bushels of yellow oats slide from the grain truck, its box hoisted high. "That's where your Cheerios come from," I told them, and saw new understanding shine in their eyes. The fields were as golden as in my memory, and the sky, the same brilliant blue. I didn't shovel grain as I did as a teenager feeling my arms growing strong, nor did I help prepare meals and drive them out to the fields as I once did. Instead, the boys and I shared a field supper with the harvest crew, during which Dainis and Jānis wangled extra helpings of dessert—lemon tarts, pumpkin pie, apple strudel—from their indulgent uncles. Then the boys ran through grain stubble with their cousins, climbed up and down the combine's steps, and played in the back of the pickup. My brothers and sister-in-law smiled, watching the children, watching the sky. Then the combine roared again.

My life here in Nova Scotia is real and true, but so was my life growing up on that Alberta farm, and I would never trade it for another. Today, however, our harvest holds no sense of urgency. Instead, it holds a sense of immediacy, of visceral connection with the earth. The sap boiling in the canning pot is sap that Vilis, the boys, and I expended time and effort to gather from our sugar maples this morning. We walked forest and parkland trails, slid down the Amphitheatre's wall, drank ice-cold maple sap laced with sweetness, and hauled the heavy load home using muscle and sinew. Our mode of harvest is basic, elemental. We see the clear sap oozing to the edge of the spile. We hear the rich plopping sound of drops hitting the surface of a nearly full sap bucket, and the tinny tap of drops splatting onto metal after the aluminum pail is emptied and reattached to the hook under the spile. We feel the weight of sap sloshing back and forth in the turquoise plastic tub as we maneuver it up the steep Amphitheatre wall and drag it home on the toboggan. In every aspect, we're fully present, fully engaged in our endeavour, embracing both the challenge and the reward. Perhaps the essence of harvest is this complete engagement, this absolute focus on the present to secure the future, whether that future is a wheat cheque or jars of maple syrup in a pantry.

Hours later, in cool evening darkness, the fire still burns. A second pot of maple syrup is almost ready to bring into the kitchen for the final boil-down. From start to finish, this has been a maple day—a day of sap and fire, smoke and syrup, and thick, sweet life.

April 16 (4°C, clear, 6:30 a.m.)

A robin calls its sharp note. As I run, I listen, hoping the note will be followed by this species' bright "cheerio-cheeriup-cheerio" song. Here in Nova Scotia, the American robin sings earlier in the morning than all other songbirds and often sings latest in the evening. In spring and summer, I'm often awakened by the sound of a robin singing solo while night is still black beyond my bedroom windows or hear one singing long past sunset. However, few robins have so far arrived from their wintering locales, and those that have aren't singing. Perhaps the weather is still too cold.

Back when I took that Ornithology class in Edmonton two decades ago, I had no special interest in birds. It's true that

I was enchanted by study skins of common loons—feathered torpedoes sleekly patterned in black-and-white, with dagger-like bills—and that during a lab I gawked in amazement at a swan's snakelike trachea. I also marvelled at the fact that a bird's respiratory tract consists of not only two lungs and associated tubes, but also of *nine* air sacs. Our avian friends don't pump air in and out of their lungs; instead, their air sacs act like bellows that push air in only one direction. The result is that their lungs are continually supplied with relatively fresh air that contains higher levels of oxygen than the mix of new and old air that enters mammalian lungs like ours. In addition to wings and light bones, avian flight requires a highly efficient respiratory system, and air sacs are the answer. So, yes, I marvelled, but I had no desire to find and identify birds in their natural setting.

Five years later and two-thirds of a continent away in Nova Scotia, I abruptly morphed into a birder. At the time I was a graduate student at Dalhousie University and lived in a second-floor flat beside Chocolate Lake, outside Halifax. The season was spring, and every day while I wrote my masters thesis at a desk beside a window, I was distracted by a bird's monotonous trill. *Just focus*, I silently told myself, yet day after day the annoying trill bored into my concentration.

Eventually my frustration mounted to the point where I realized that the only way to restore my equilibrium was to identify the culprit. To that end, I borrowed Vilis's binoculars and bird guide and scanned the yard surrounding our flat, although I had no idea what size, shape, or colour of bird I was searching for. Over a period of four days, I prowled the yard and nearby woods for hours, determined to locate and identify the annoying songster. At last, I spotted a tiny grey-and-white shape belting out the nauseatingly familiar trill from the topmost twig of a lofty evergreen tree. Desperately, I flipped through the bird guide and found the triller—a dark-eyed junco. In that moment of discovery, a new world opened to me, a world of mystery voices, fleeting glimpses, and thrilling definitive sightings. Birding cast its spell over me, and I've been lost in it ever since.

I run in silence broken only by the crunch of ice beneath my sneakers. Still no robin's song. When I reach Balmoral Road, the sun splits me into gold and grey, as if I were the earth, half of me bathed in warm light and the other half in shadow. Eight minutes ago, these golden rays of energy left the sun's surface, spewed outward as hydrogen fusion's radiant breath of heat and light. Contented, I turn and run home. I don't have the robin's song, but I have the sun, basis of all life on this planet. As our Northern Hemisphere days lengthen and warm beneath Sol's rays, the robin's urge to proclaim its territory will grow until once again *Turdus migratorius's* clear ebullient melody will wake me when night is still black beyond my bedroom windows.

Snow melts. Soil dries. Matheson Brook runs strong now, its progress no longer impeded by plates of ice thrust over each other, under each other, or braced against each other with the strength of winter. Grey and dirty, some ice plates remain, hung on the brook's rim by flood waters. The fact that a few of these plates rest two metres higher than the current water level is a silent testament to the strength of this brook in which my family walks and plays in summer.

The sun is setting, and I'm en route to our zebra mailbox, with its white plastic teeth and plum-painted tonsils, on Matheson Brook Road. I cross our old wooden bridge that has sagged ever lower during our nearly six years at Ravenhill. Each winter, the decrepit structure acts as a blockade that causes ice to pile into ridges upstream of it, and each winter, during rainstorms, those ice jams invariably cause flooding. Then Matheson Brook—swollen with run-off and diverted from its usual path—pours onto the meadow adjacent to its stream bed, fills to overflowing a small pond at the base of our front road hill, and rushes over the front road as the water relentlessly seeks a path to the ocean. At present, however, with most of the brook ice melted, washed away, or stranded on shore, that flood scenario is in the past for this year. As I near the end of our front road, I catch the startling silver shimmer of sunlight on moving water and know that truly the brook is free.

After collecting mail, I retrace my steps, pausing to check serviceberry, alder, birch, and maple buds for signs of opening, but no, they still sleep. Raspberry canes arc in gentle curves beside the road, and blackberry canes, in fiercer curves. They, too, still sleep. I pause again, this time beside the small pond created by seepage down the hill and high water on the brook. This evening, the pond's entire surface wears a mottled layer of silt several centimetres thick on melting ice. The silt was deposited by the last flood. How will it affect overwintering amphibians and insects—green frogs and red-spotted newts, dragonfly nymphs and caddis fly larvae—that inhabit the pond? Will the silt settle on these aquatic creatures with the silence of suffocation, or will they simply crawl through it to flourish in a richer environment pulsing with new nutrients, like the post-flood banks of the Nile before the Aswan High Dam was built? I suspect the latter. If the former were true, our pond would be barren each spring.

I stride up the hill, out of the brook valley's cool shadows and away from the water's rich vowels of passage over large rocks and its incessant riffling over pebbles. On the hill, alder catkins' brownish covers are split, revealing spring's green promise. Atop the hill, I tally our nine chickens and close the coop—a long, boxy building with a slanted roof that Vilis and I built four years ago using tamarack poles from our land and rough-milled spruce boards. In the darkening evening, a robin calls a single note, but again I hear no song.

Light fades. The sky turns deep blue, and I know stars will shine in all their clear beauty tonight. Orion, the great hunter, will pursue Lepus, the hare, with his dogs, Canis Major and Canis Minor, his shield raised against Taurus, the bull. The Big Dipper will point to Polaris, the North Star, and beyond Polaris, Cassiopeia, the queen. And this side of Cassiopeia, a rare sight in the night sky will glow like an incandescent ball of fire tossed through the darkness, its two tails gleaming like the glistening sails of a tall ship tacking hard against a solar wind. Astronomers call the celestial phenomenon Comet Hale-Bopp and describe its blue tail as composed of charged gas molecules reclaiming electrons, and its yellowish tail as composed of dust blown off the comet's core.[7] I don't doubt these descriptions, but when a clear spring night triggers astronomical anticipation, I say Comet Hale-Bopp is a ship, a tall ship, the most beautiful ship that will ever in my lifetime sail the night seas. In the deepest hours of the night, I'll rise from my bed, shrug on a jacket, and step, tingling, into the night to observe her passage.

April 20 (early morning)

Dark blue clouds scud across the morning sky while evening grosbeaks call from a young elm tree at the edge of our yard. I take these sensory images with me as I run the back road through the meadows south of the house. At the edge of the woods, small circular holes—some larger, some smaller—pock the snow, each one an impact crater formed by a raindrop that fell through the tree canopy. I set my feet one ahead of the other on the rain-pocked snow. Some days the only reward for running is the open road ahead. Those days hold no surprises—no barred owl flying overhead, no fisher bounding across the path and turning to stare at me, no raccoon family at the forest edge. Yet I know that the unexpected would not be unexpected if seen all the time. I reach Balmoral Road and run home, my footsteps slow and heavy. I saw nothing new, nothing unusual that struck a spark in me.

In the house I remove my sneakers and ponder unexpected happenings. Two days ago, I glanced up from kneading bread and saw three white-tailed deer strolling through a meadow south of the house. That sight was unexpected. Last week, I walked past the kitchen patio doors and was startled when I noticed what appeared to be soil bouncing in the front flowerbed. Upon closer examination, I discerned a throng of juncos and sparrows, their dark-coloured heads bobbing up and down as they pecked for fallen seeds. The birds' repeated bobbing action created the bouncing effect, which I'd never seen before. Lastly, this morning while running home, although I observed nothing unusual in the world around me, I was struck by an odd incongruence that hinted at something more profound than the physical reality. Nearly all my footprints on the back road were silver-grey on thin snow, but one was brown, mud-stained—startling in its nonconformity. It was as if that one step, although set in a line of others, walked a different path.

Now I think that perhaps that step is me. Being a writer is inherently a different path, a life of solitary pursuit. Robert Frost penned the classic poem "The Road Not Taken," which is all about choosing a different path in a yellow wood and choosing a different path in life.

I'm a patient woman. I've gifted myself a year to explore, reflect, and write about the forest and meadows on my doorstep. During the seasons that come, I'll seek revelation while running and pursue insight while walking. I'll sit amid grasses, still and silent, essaying discernment. And I'll wait, confident that out of nowhere the unexpected will leap at my senses, steal my complacency, and catapult me into fresh awareness of all that surrounds me.

April 22 (7°C, cloud/clear, 10:40 a.m.)

I snip off a five-centimetre sprout of chives and revel in its fresh tang. Rhubarb, tulips, and daffodils have also sprouted, uncovered since yesterday's sunshine melted the last of a three-days-ago dump of wet snow. So far in April, we've had two such snowstorms. Will there be more? I gather forks lost in grass during autumn barbecues, then lie on our picnic table and stare at the sky. Clouds above me are lobed brains, but those in the west are mares' tails; those in the east, carded wool. Our male cat, Charcoal, bounds lightly onto the table and rubs his tail in my face. He's much friendlier now than he was before a six-month disappearance last year. We thought him dead and then rejoiced at his appearance on the porch one day last fall, like a prodigal son gone and returned. For a short time, three cats prowled our domain; then Dusty—our sleek, green-eyed Dusty—disappeared.

I sit up and see the work of our hands—asparagus patch, domestic blackberries, shrub apricot, Saskatoon berries, rhubarb, high-bush blueberries, soil awaiting strawberries, fire pit, compost heap. If I were to turn a half-circle, I could view stacked firewood, the chicken coop, and awakening herbs. Ravenhill, with its rich soil, abundant wildlife, and solitude, is a good place for a pair of biologists to make a home and raise a family.

Even as I have this thought, Samson crows, and in almost the same instant a dark-eyed junco trills its song. Once when Vilis and I lived at a scientific research station beside Kluane Lake in the Yukon, I heard almost simultaneously a rooster's crow and a loon's wild cry. The two sounds struck me as being worlds apart, yet domestic and wild coexisted there. I wish the same for Ravenhill. My dream is to bring new life to our century-and-a-half-old house and ancient apple trees yet have them blend in with nature. Conservation pioneer Aldo Leopold wrote, "There are some who can live without wild things, and some who cannot."[8] Like him, I count myself among those who cannot. Here at Ravenhill, each day our brook, meadows, and forest spark light into my soul.

Charcoal jumps down from the picnic table and chews greening grass. I hear the brook rushing at the bottom of the hill, and wings flitting in the yard. Above me, clouds pile into white-sailed schooners that race from west to east on cool spring winds.

April 27 (-9°C, clear, calm, 6:45 a.m.)

This morning, frost lies cold and white on the ground as I release chickens into the yard. For the first time this year, a white-throated sparrow's whistled "oh, sweet Canada-Canada-Canada" and a ruby-crowned kinglet's rollicking melody brighten the air. Both songbirds are short-distance migrants that spend our Canadian winter in the United States or Mexico and fly north in spring in pursuit of abundant food to meet the energy demands of breeding and of raising young.

Bird migration is an epic story within the vast encyclopedia of animal natural history. Why do tiny ruby-throated hummingbirds that nest in my yard undertake migration journeys of thousands of kilometres to and from their overwintering grounds in Central America? Why do arctic terns, like those who dive-bombed me while I walked a shoreline on Ellesmere Island, raise their offspring on High Arctic tundra, then fly south, deep into the Southern Hemisphere to overwinter?

Of course, it's all about food. Reproduction places a monumental energy demand on birds, as it does other organisms; yet, most of our avian friends have a big advantage over earthbound creatures—they can fly. Not limited to terrestrial movement, they're free to travel "as the crow flies" to new, lusher feeding grounds as seasons change. When temperatures warm during our Northern Hemisphere spring and summer, abundant blossoms offer nectar to hummingbirds, and farther north, explosions of insects and an abundance of small fish feed arctic terns. When the blossoms and insects are killed by frost, and sea ice forms above the fish, it's time to for hummers and terns to head south to milder temperatures and their winter food supplies. The graceful arctic tern, with its long, angular wings, is the world champion of bird migrators, flying an astonishing 40,000 kilometres from its northern breeding grounds to summer in Antarctica—essentially from pole to pole.

Energized by birdsong and reluctant to return to the house, I remove old flower stalks from the flowerbed in the front yard before hanging laundry. Frozen soil is hard and lumpy underfoot, and when I rub frost from the clothesline, white flakes shoot out like sawdust exploding from a saw blade.

"Breakfast's ready!" Vilis calls, so I leave the world beyond walls for the one within walls. While we pour maple syrup onto our pancakes, a yearling rooster—the one we call Young Samson due to his resemblance to his sire—displays to a pullet just beyond the edge of our porch. The young hen pecks him, sending him on his way. Dainis and Jānis laugh. Inspired by the hen's stand-offish behaviour, they name her Fusso.

We first raised chickens four years ago, starting with thirty-five Meat Kings and six Golden Comet layers that we bought as day-old chicks. The tiny chirping fluff balls enchanted us. Initially, we kept them in our house's L-shaped porch, which at that time was walled in. We fed and watered the wee birds, changed their newspaper bedding, and kept them curtained away from my dog, Skyree. However, as the chicks grew, so did the volume and stink of their excrement until our noses demanded we move them elsewhere.

An obstacle stood in the way of the move, and it was a big one. The coop was only partly built. Vilis and I hummed and hawed about what to do and then dragged an old metal livestock-watering tank—tall-sided and shaped like a thick sausage—into the partially-constructed coop. We gathered dry grass for bedding and placed the chicks' feeders, waterers, and grit on the bedding inside the tank. We set up heat lamps and, after rigging up a roof made from several plastic tarpaulins, carefully transferred the chicks into their temporary shelter.

That spring was cold, wet, and windy, hindering coop construction. As a result, the chicks occupied their makeshift accommodation longer than we intended. Finally, on a sunny morning when the coop was almost finished, my sons and I introduced the chicks to their new home. We first spread a thick bed of wood shavings over the coop's floor and moved the feeders and waterers out from the tank. Five-year-old Dainis climbed inside the tank and gently tossed chicks onto the wood shavings. I reached inside the tank and handed chicks to two-year-old Jānis, so he also could release them. At first the young birds simply stood motionless, uncertain what to do in the large, unfamiliar enclosure. Then they began to walk about quickly and broke into brief dashes. They chirped excitedly and leapt into the air with their necks and wings outstretched, as though embracing the sudden luxury of space. Sunlight streaming through an open doorway painted the new walls and wood shavings golden, the leaping bodies of the Meat Kings pure brilliant white, and in that instant, I knew that even chickens can feel joy.

We can never predict where we'll find enlightenment. Even when we're occupied in the most mundane task, such as releasing chicks into a new space, a fleeting image might arrest us. In that instant, the scene before us is greater than the sum of its parts. It transports us to a higher level of awareness, both of the intricacies of the world in which we live, and of ourselves. The intensity of the moment, rich with revelation, leaves us changed. I hope to carry the image of that day releasing the chicks in the coop to my last breath.

That first year we raised chickens, we butchered the Meat Kings in late July and early August—thirty-one of the original thirty-five—and kept a rooster to crow in the mornings. Dainis and Jānis named him Old White. Heavy, clumsy, and vicious, Old White lasted until May of the following year, when he keeled over one morning and fell from the roost amid much squawking from the hens. "Old White kicked the bucket!" little Jāni told the foundation crew demolishing our house's chimney in preparation for installing a cement basement. The men grinned, slapped their thighs, and laughingly repeated the breathless news.

That summer, friends brought us gifts of a rooster and a feisty cinnamon-coloured hen that Dainis named Colour-Cluck. My first sight of the rooster stopped me in my tracks—shimmering rust-gold neck, iridescent black-green belly and tail, neck arched high, feet stepping out boldly, with each leg backed by an eight-centimetre spur. "Now *that's* a rooster!" I declared and promptly named him Samson.

Young Samson is the image of his father. Beyond the porch, he fluffs up his rust-gold neck feathers and dances around Fusso, bowing and scraping his black wings on the ground. He is not, however, the only contender for the pullet's affections. Samson's other son, a black-and-white-speckled rooster, runs to challenge his brother, and the fight is on. Samson jumps a metre into the air, with spurs out, meeting his brother mid-air. The two cocks peck and slash at each other. Their neck feathers—erected to intimidate their opponent—look like feather-bowls holding their heads. Repeatedly, the roosters spring into the air, spur-stabbing until the speckled rooster gives in and makes a break for the spruce windbreak, with Young Samson in hot pursuit.

Our quarreling roosters remind me of male sharp-tailed grouse that I observed for two weeks one spring in southwestern Manitoba. The year was 1984. At the time, I was assisting Vilis with his doctoral research into the hunting habits of striped skunks; however, I also occasionally helped another graduate student, Michael Gratson, with his studies of the effects of predation on the nesting success of sharp-tailed grouse.

Each spring in southern Manitoba, male sharp-tailed grouse ("chickens" to locals) congregate on open patches of ground, such as pasture or undeveloped prairie. These sites, known as dancing grounds, are used by male sharp-tails year after year, making it easy to locate and observe their mating activities. The males in action look almost ludicrous, like wind-up toys. With wings bowed and purple neck-sacs inflated, they stamp out a ritualized dance while rattling their wing quills to produce a whirring sound. Each male is highly territorial and will attack any other male that dares to enter its territory—rather like Young Samson attacked his brother. The entire group of male grouse displaying on a particular dancing ground, as well as the dancing ground itself, is referred to by scientists as a lek. Female sharp-tails, enticed by the males' dancing and calling, seek mates on the lek.

During the two weeks that I observed the sharp-tails from a blind, my pencil flashed ever more quickly in my field notebook as more and more hens sought the dancing ground, and as the intensity of the males' dancing and fighting escalated to the point where I could barely keep up with it. Picture the following scenario, which I attempted to document one morning. Ten males whirl out their dance, rattle their quills, and cackle in a frenzied display intended to demonstrate their fitness as potential mates. Meanwhile, nine hens stroll casually among the hormone-fueled males, assessing their prowess. Vicious battles break out between males at invisible (to me) territorial borders. During such disputes, other males sneak onto the territories of the fighting males. A few males attempt to copulate with hens, only to have their prospective mates simply walk out from under them. Other males mount hens and are knocked off by competing males. Still other males mount hens and, like bareback riders at a rodeo, hang on for the required seconds. Then the males drop off the hens and get back to the business of defending their territories while their mates smooth ruffled feathers and fly or sedately walk from the lek.

Each morning the mating extravaganza started at first light and lasted roughly two hours. And then, as though a switch had been flicked, hens deserted the lek, and the male sharp-tails settled into pairs of ground-hugging lumps that stared face-to-face at each other across territorial borders and occasionally uttered a muted "too…too…too-*kloo*."

During the afternoon, our young roosters continue to flaunt their handsomeness to the hens and fight with each other. Birds of a different sort—black-capped chickadees—flutter around me while I trim a honeysuckle in our front yard flowerbed. It's a novel sensation to hear the chickadees' tiny beaks tap against the feeder bottom as they pick up seeds, and to feel their wings' whirr close to my head. Today is truly a *large* day.

May 5 (3°C, clear, 8:20 a.m.)

This morning the speckled rooster and Fusso call loudly from the coop as I carry a laundry basket behind the house and hang wet clothing on the clothesline, careful to spread each item wide to catch the sun's rays and the whipping wind. Evaporation dries my wash, lifts sweat from hot faces, wilts the broad leaves of squashes in summer, and on warm winter days forms clouds of fog over sublimating snow as frozen moisture jumps from solid crystals to gaseous vapour, bypassing the liquid state between—the puddles and trickle-streams.

One day last spring, Dainis, Jānis, and I did an evaporation experiment. We took four cloth napkins, soaked them in water, and wrung them out. We spread one in the sun at the porch's edge and one in the shade. We balled up the other two napkins and set one beside each spread-out napkin. "Which do you think will dry fastest?" I asked, to which both my sons replied, "The ones in the sun." We waited, and sure enough, the spread-out napkin in the sun dried fastest; however, the balled-up napkin in the sun dried more slowly than the spread-out napkin in the shade. To no one's surprise, the balled-up napkin in the shade dried slowest of all. I explained that the rate of evaporation depends on the amount of surface area exposed to air, sun, and wind. Each time my sons leave wet towels in rumpled heaps on the bathroom floor, I remind them of that experiment.

Now I stand beyond the clothesline and garden, with no man-made structures ahead of me. Blackberry canes, dead and alive, arc in shallow curves like thorny ribs emerging from the earth. Charcoal rests nearby in dry grass, licking his forepaws and shoulder before melting into cat repose. I stroll into the parkland back of the garden, and Charcoal rises from his rest to accompany me. I voice a soft "pssh, pssh" to a yellow-rumped warbler flitting about in trees beside me. The handsome male perches in a spruce tree only three metres away, allowing me to admire its grey-white-black plumage highlighted with yellow. A second yellow-rump lands in a nearby tree, as do a black-capped chickadee and a dark-eyed junco. Are these songbirds as intrigued by me as I am by them? Or are they cat-watching from the safety of trees and open space where they can flee?

Sometimes it feels strange to go walking with a cat, yet as a child growing up on an Alberta farm, I walked with my family on a dusty path through oats or wheat to check cattle in the pasture beyond, and our cats often accompanied us. Tails high, they padded beside us, their paws falling silently into warm grey dust, the same fine dust that puffed up between my bare toes. If my father were still alive, I would thank him for those walks, for that farm, for the life he gave me. We are always where we came from, at least some part of us is. In this moment, all of me is.

Now Charcoal is stalking, his tail twitching. The warblers, chickadee, and junco fly to higher perches and safety. They don't like the company I keep.

Cold wind knifes through my jacket, as sharp and biting as the unexpected grief and homesickness in my heart. How is it that memories can come upon us so suddenly and have such power to overwhelm us, even though decades might have passed since the experiences that created those memories? Here I stand, vulnerable to the past, grieving for a father I lost nearly twenty years ago, and homesick for a place I might not choose to live in now. Each time I visit the farming community where I grew up, it seems tame and cultivated, yet the people of that farmland—my people—arc like ribs from the prairie dust, many of them thorny with pride, all of them sturdy. If my country were a living being, some Canadians like me might see the North as its spirit, the East Coast and West Coast, its reaching hands, the central cities, its brain, and the prairies, its heart. Perhaps the reason for my sudden grief and nostalgia is that I left that heart behind. My heritage, my family, and my father's grave are five provinces and thousands of kilometres away.

So. Wonder is not the only emotion to strike suddenly. Grief and homesickness can also take us like a flash flood that washes away sandbags we've consciously or subconsciously stacked high. Buried but not obscured by time, tears and sorrow lie quiescent until a trigger jolts them into leaping pain. As I discover this morning, that trigger might be as innocuous as a cat providing company on a birding walk in the woods.

Cold stings my hands and feet. Too much thinking and not enough walking. *I'm coming, Dad.*

May 6 (8°C, clear, strong NW wind, 10:45 a.m.)

The chickens are cautious, staying close to the porch and spruce windbreak under which they can escape in a flash. They have good reason to be cautious. Earlier this morning while Vilis, Dainis, Jānis, and I poured maple syrup onto our Cream of Wheat, a northern goshawk skimmed across the yard, perched for a moment on the chicken yard door, and then took aim at the speckled rooster and flew. Startled by the attack, my husband, sons, and I jumped up from the table and watched while the goshawk closed in on the rooster as he sprinted for the windbreak. Less than a foot of air separated the hawk's talons from the rooster's head when my sons and I burst onto the porch, our yells announcing that reinforcements had arrived. The goshawk braked, lifted gracefully, and flew to the west.

"Well, it would have solved our rooster problem," Vilis said, meaning that a successful kill by the goshawk would have ended the fights between Young Samson and his brother.

True, but I still have enough farmer in me to protect my livestock from even the most beautiful of predators. I believe this goshawk is the one that killed a young rooster behind the coop last November and returned morning and evening to feed on it. As a family, we crept silently toward the kill site and hid behind the woodpile while peering out to enjoy the adult hawk's red eyes and slate-coloured back, wings, and tail. Perhaps this *Accipiter gentilis* is also the hawk that, as an immature bird, razed our chickens one spring. It might even be the goshawk that killed a Leghorn hen right under our living-room window two autumns ago. Reading on the sofa, the boys and I heard loud chicken squawks followed, seconds later, by white feathers floating up past the window.

Was this Colour-Cluck's fate? We returned home later than anticipated one evening last week, long after dark, and found three chickens missing from the still-open coop: Samson, Young Samson, and Colour-Cluck. We found Young Samson wandering about the yard in the dark, and I spotted Samson near the spruce windbreak the next morning, but Colour-Cluck—our spirited, blue-egg-laying, real-chicken-sounding Colour-Cluck—was gone. Here at Ravenhill, we've learned that raising livestock in predator territory means you inevitably pay a tax. But not this morning.

After breakfast, I step onto the porch and hear a fox barking. Today is a day for predators. I call Vilis and the boys, and we listen to the fox's raspy voice echoing from the lower meadow beside the brook. Dainis runs for binoculars. Vilis and I share a grin. There's something about knowing that a predator is near that quickens the blood and sharpens the mind, that puts you back in the forest or on the tundra, knowing that danger is near, that within your sphere is an animal that kills to live. Occasionally, that animal might even have the power to kill you.

Let me tell you a story. As a graduate student I spent two summers, in 1980 and 1981, studying plants on Ellesmere

Island in Canada's High Arctic. The research site was a lowland oasis of lush (for the High Arctic) plant life within an otherwise barren landscape of mountains and glaciers classified by ecologists as polar desert. An abandoned Royal Canadian Mounted Police station on the shore of Alexandra Fiord, half-way up Ellesmere Island's east coast and about forty kilometres from Greenland as the raven flies, served as our research base. Each day I filled a pack with research gear and hiked to one or more of my field sites on the treeless tundra.

One day, weary from measuring new growth of arctic white heather—the subject of my research—I took a break and began counting ringed seals on the ice-covered fiord in the distance. Through my binoculars, I slowly worked my way along the fiord backed by glacier-hung mountains while I tallied dark streamlined shapes basking beside their breathing holes. I shifted my view a little farther toward the head of the fiord—and my body locked in stillness. Another animal was on the ice. My pulse began to race, and life seemed suddenly balanced on a knife edge as I watched a polar bear lift a ringed seal in its jaws, gripping its prey by the neck. The bear's fur was yellowed. The wet seal was almost black. White ice surrounding bear and seal was smudged with old grey ice and smears of shining water. The bear shook that seal as a cat shakes a rat. Heart pounding, I packed my field gear and hiked back to base camp on the fiord's shore, where I alerted the other half-dozen researchers to the bear's presence. Several who had been sleeping in tents at night moved indoors, but other than that, we continued our work, unarmed, albeit with increased vigilance.

Two days after the first bear sighting, my dog, Skyree, barked incessantly during the evening, an unusual behaviour. Taking his cue, I scanned the ice and again spotted a polar bear, perhaps the same one. It crawled on its belly toward a seal resting beside a breathing hole but met with no hunting success. The alert seal detected its arch enemy and slid into the water. The bear abandoned the hunt, bounded to a rocky island, and as if playing, raced up and down craggy cliffs before loping in huge, ice-eating strides toward the head of the fiord. On other occasions, wind blew inland from the fiord, causing Skyree to bark a warning, but I never saw the bear again and have not since observed such magnificent power exhibited by a single animal.

Wild country tests us, enchants us, invigorates us, inspires us, and sometimes terrifies us. It brings us face to face with our mortality and reveals the depth of our character in ways no town or city ever could. On my first day at Alexandra Fiord, I put away my money, credit card, and keys and didn't touch them again for months, until packing to leave. Necessities within civilization, cash, keys, and credit were irrelevant on that isolated High Arctic lowland. There, my money was the health of my body, my credit card, the keenness of my brain, and my keys, the strength of my spirit. At the end of each of those two summers, I flew back to Halifax and was shocked by the lushness of hydrangea blossoms, was afraid to cross busy streets, and shrank from concrete towers and throngs of people. But I was stronger inside, for I'd watched a polar bear shake a seal in its jaws, had known it could do the same to me, and had still walked on the tundra.

Northern Nova Scotia isn't polar bear country, but we've had visits from at least one black bear that left droppings on the old wood road here at Ravenhill. As well, that bear, or another, raced across our bridge one spring afternoon when I surprised it by driving our car down the hill. Again, my blood quickened, and in the back seat, Dainis and Jānis shouted, their voices flaming with excitement. Edward Abbey wrote, "...wilderness is not a luxury but a necessity of the human spirit, and as vital to our lives as water and good bread."[9] Our seventy-eight acres are not wilderness, but much of our land is reverting to *wildness* after more than a century of domestication. And that's what we want. Good bread. Good water. Here, our spirits sing.

May 10 (1°C, overcast with rain or snow, strong NW wind, 9 a.m.)

Dainis is planting trees with the 1st Tatamagouche Cub pack, and I'm worried. The temperature is one degree above freezing, and a strong wind blasts out of the northwest. Snow or rain is falling, depending on which time I look out the window. This is hypothermia weather. Unable to allay my concern, I take Dainis his insulated winter boots to replace his rubber boots and give him my pile-lined gloves to replace his soggy ones. Then I stay and help plant trees, pressing conifer seedling plugs into a grassy field.

Each seedling, if it survives, will become part of the earth's lungs, taking in carbon dioxide and releasing oxygen through the process of photosynthesis. Some of the seedlings seem almost too small, too spindly to hold the power for such tremendous life-giving benefit—creating oxygen—yet I'm reminded that similar latency abounds in other growing things. When I cup bean seeds in my hand, I hold the key to capturing atmospheric nitrogen and adding it to the soil in forms that can be used by growing plants. When I plant a tiny chive seed, it will eventually produce nectar-rich flowers that will nourish bees and butterflies, which are valuable pollinators. Nature gives us the potential to do great good for the earth, and in doing good for the earth, we do good for ourselves.

Other tree planters are shadowy figures moving among rows of small red, blue, or orange flags screened by white. At times, the snow falls so thickly I can barely see the end of the field. Based on our years at Ravenhill, I've learned that here in northern Nova Scotia May is a battleground where winter and spring clash until summer arrives overnight.

May 15 (6°C, clear, S wind, 7 a.m.)

This is a day of spring, of frog song and drying winds, of clear sunlight and awakening twigs. A fresh breeze filled with new scents sparks a familiar desire in me. If I could, I'd escape. I'd leave behind teaching and housework and motherhood and instead hike a mountain or canoe a swirling river. Every spring, Vilis and I experience this same desire to leave our current lives behind and embark on a new adventure. We breathe spring's evocative air and yearn to travel north or west to wilder, less settled country, to see mountains rise against the sky or to walk on tundra.

A spring exodus to wild country was the story of our lives when we were university students. Vilis measured songbird territories in northern Ontario. I collected fish fossils in the

Alberta foothills. He researched the hunting behaviour of ermine in the Northwest Territories. I trapped small mammals at the same site, and we fell in love. I flew to Ellesmere Island to study High Arctic plants. He flew to Alberta's tar sands country to assist a mathematical ecologist. After marrying, we moved from Halifax to Victoria, where Vilis began his doctoral research; however, spring took us to southwestern Manitoba for his field studies. When his research finished, a snowshoe hare project lured us, yet again in spring, to the Yukon and a cabin beside Kluane Lake. It was during those nomadic years that my brother Tod christened us "the gypsies."

Kids changed all that. Parenthood made us look carefully in our wallets. It made us want to have a home.

Life consists of different seasons. While immersed in one season, we might occasionally yearn for another, as I do this morning. However, no number of adventures in far-flung wild places could replace the richness of these years with Dainis and Jānis. Here at Ravenhill, our outdoor adventures are smaller, usually close to home, and often educational in scope as Vilis and I seek to awaken in our sons a love for the outdoors and a desire to experience nature often—to hike and camp and paddle and ski. When Dainis and Jānis, now nine and six, are eighteen, they'll be the ones seeking adventure, they'll be the ones leaving home and parents to find their own future and explore the world. Then Vilis and I will be free to set out on our own again. In her song "The Circle Game," Joni Mitchell sings of life's seasons going round and round, and of not being able to return to a previous season, only to look back at it. This morning, spring's sweet air sends my thoughts tumbling back through time even as I cherish the present.

Charcoal meows and comes for petting, his neck, thick and muscular beneath my hand, his dark fur, glossy in the sunlight. I wonder if he ever thinks of escape, of leaving this peaceful yard for rarer air and less trodden soil. Then I recall that Charcoal disappeared for six months last year and returned to Ravenhill a friendlier cat. Perhaps new adventures should always lead us to treasure more deeply what we already have.

May 18 (6°C, cloudy, 5:30 p.m.)

It's late afternoon after rain, and I'm walking to our family campsite near the Amphitheatre. My arms are filled with fresh balsam fir and hemlock boughs intended to cushion our sleeping bags during a night of spring camping. The sweet, woodsy scent of fir hangs in the air, and all around me, raindrops glisten on trees. Like a camera lens, each shining water drop reflects the world upside down. I take another step forward and notice a deep-blue light gleaming in the forest ahead of me. The sight of it halts my steps. I've never seen a blue light like this in our forest, that is, not until now.

With breath held, I take slow steps forward, testing the light. As I draw closer, it turns from blue to yellow, then to red, and then to a clear raindrop. At the age of forty-one, I'm observing a natural phenomenon I've never seen before—a drop prism, a raindrop refracting light into colours of the visible spectrum. Mesmerized, I slowly back up. As I do, a red light replaces the clear drop and is, in turn, replaced by a yellow light, then a green light, and then, once more, the vivid deep-blue light. I glance to either side of the blue light and see, toward the west, other raindrops caught in the sun's rays. A scintillating bevy of rainbow colours flashes from the tips of twigs.

A month ago, I sat in my kitchen pondering unexpected happenings. Now here I stand in a greening wood, and the unexpected I so wished for is here, leaping at my senses, stealing my complacency, and catapulting me into fresh awareness of all that surrounds me. Laughter races through me, as though I gleefully share an insider joke with the Creator. I count it a privilege simply to exist in the presence of these exquisite drop prisms. Annie Dillard wrote, "Our life is a faint tracing on the surface of mystery."[10] Surrounded by fragments of liquid rainbow, today I'm dancing on that mystery.

When I reach our campsite, I deposit my armful of boughs on a bed of old fir branches that is still surprisingly resilient. Two years ago, in early March, my family laid freshly cut branches here on snow to create a springy mattress for a night of winter camping. The boys were three and six at the time and keen to spend a night in the outdoors. We skied from the house, with Vilis pulling our toboggan loaded with bedding, food, and camping equipment. After reaching this evergreen-circled clearing, we braced lean-to poles between two sturdy spruce trees and covered the poles with orange plastic tarpaulins. Then we fashioned a mattress of balsam fir boughs under the lean-to and laid our sleeping bags on it.

Darkness came early that night. We cooked our supper over a campfire by firelight. Above us, stars gleamed like white points of astral fire set against deep blue sky. We became part of the night—the snow and sky, the clearing and campfire. It was cold, the temperature dipping to -20°C. Vilis and I sandwiched our sons between us and piled blankets on top of them as we all huddled deep in our sleeping bags. During the night, Vilis fed the fire every two hours, becoming chilled to the bone, his boots completely frozen. I also felt the frigid night's bite, but Dainis and Jānis didn't. Snug and warm within a stack of boughs and blankets, each woke smiling, wanting to stay another night.

Now, as I collect more boughs, Dainis runs to me, excitement blazing in his eyes. "Magi, have you seen the coloured drops?"

"Yes!"

He and I exclaim over red, yellow, and blue drops ahead, toward the sun, and point them out to Vilis and Jānis. Then we all marvel at long white fingers of mist sliding through the forest surrounding our campsite. As we arrange new boughs over old, chickadees flit secretively in the trees.

May 21 (5°C, overcast, calm, wet from rain, 8:20 a.m.)

This morning the speckled rooster walks alone, ostracized. Two days ago, he and Young Samson almost killed each other. I had to kick them apart, and still they wanted more although their beaks were sticky with bloodied feathers and opened wide and panting from all the chasing, jumping, and kicking. Vilis and I have decided to butcher Young Samson and see how Samson and his speckled son get along. Dainis and Jānis object to this plan, but how else can we deal with the combatants? The question is, when to do the deed?

A blue jay flies to an apple tree in the yard and then downhill toward Matheson Brook, its iridescent blue back and cerulean wings and tail a brilliant flash against the sky. In spruce forest to the west of the yard, an ovenbird's brisk "te-*cher*, te-*cher*, te-*cher*" song snaps against the morning air. I see the blue jay, but not the ovenbird. The latter reveals its identity with its voice.

With fifteen years of birding under my belt, I now bird as much, if not more, by sound as by sight. I quickly learned that the leafing out of deciduous trees in late spring coincides with the most intense period of breeding activity for migratory songbirds. Once leaves hinder the view, getting a clear look at an avian songster in a tree can be a difficult and time-consuming task. Leaves, however, don't block sound. Each spring I review the songs of wood-warblers, flycatchers, and other migrants to re-acquaint myself with their melodies. Consider the ovenbird proclaiming its presence right now. An ovenbird is a rather drab cousin of other more vibrantly hued wood-warblers, such as the flame-throated blackburnian warbler, the buttery yellow warbler, and the striking black-and-red American redstart. The colour of faded leaves, an ovenbird easily blends in with tree bark and the forest floor, and it's a skulker, which makes it doubly challenging to spot. The ovenbird's song, however, is loud and unmistakable, and once learned, will never be forgotten.

My mind takes me back to a time when my interest in birds played an intriguing role during a profound life experience. Nine years ago, as I lay in Whitehorse General Hospital awaiting the birth of my first child, I watched cliff swallows and belted kingfishers fly beyond the hospital window and heard the songs of dark-eyed juncos, robins, and a yellow-rumped warbler. The flitting wings and avian voices grounded me and connected me with the world beyond my own drama.

They were a light in the darkness of pain and uncertainty, a flicker of joy and normalcy within the anxiety and suffering that accompanied bringing a new life into the world. Now, my mind jumps forward to the opposite end of the life spectrum, and I believe I would like to die with birdsong in my ears.

At the edge of our yard, new alder leaves are crimped ovals of a green so lush it makes me want to stuff handfuls in my mouth; instead, I feed my eyes. Raindrops bead heart-shaped aspen leaves trembling on red petioles, and a spider's web weighted with silver-edged raindrops sags onto grass. Where is the web's owner? What is the likelihood of capturing prey on a cool, sunless day like this? But then, is there any normal hunting day for a spider? In the musical production *The Addams Family*, ghoulish Morticia tells her daughter Wednesday, "Normal is an illusion, darling. What's normal for the spider is a calamity for the fly."[11] This morning, I see no flies caught in calamity, yet I'm reminded that this web I observe purely as a matter of interest does not exist to satisfy my curiosity. It's a death-trap, the setting for a natural drama in which one organism will die and feed another. If no human were ever to see it, this web would still exist and serve its function. The earth and its creatures don't exist for the pleasure of humans. They exist for the pleasure of God.

I stroll to our front flowerbed, where short blade-like daffodil leaves protrude from the soil. So far, spring has brought little warmth to raise them high. Last year, days earlier than this, my family took a full bouquet of opened tulips to a nearby friend. To date this year, we have no open tulips. In garden beds west of the house, rhubarb and asparagus are also slow in coming, and my weedy vegetable garden behind the house is yet to be tilled. As always, the weather dictates garden schedules, and this year it commands patience. I return to the asparagus patch, snap off a handful of spears, and take their weak promise of more to my kitchen.

May 22 (6°C, clear, 8:45 a.m.)

After days of rain, sunrise under a clear sky pours golden light over Ravenhill, painting tree bark with stark detail and filling me with optimism. I'm excited, nervous. As a freelance writer, I teach creative writing workshops in schools, facilitated by the Writers Federation of Nova Scotia. I've visited schools from Halifax to Lockeport on Nova Scotia's South Shore, from Berwick to Weymouth on the Fundy side of the province, and from Nappan to Saint Andrews in the north. This morning I'll drive to East Pictou Rural High School in Thorburn and spend the day leading Grade 11 students in creative writing exercises. I have little experience working with high school students. Most of my visits are to elementary schools, where I show slides of wildlife, read from articles I've written for the children's nature magazine, *Ranger Rick*, and answer countless questions about wildlife and writing. I return home from those visits with my voice tired, but my mood buoyant. Today will present more complex challenges. I wonder how it will go.

In late afternoon, I'm drained, relieved to return to Ravenhill's solitude. The students had so few questions, so few answers. Who stole their curiosity? What left them staring at me blankly? Vilis, who teaches and does research at the Nova Scotia

Agricultural College, refers to students who show little or no apparent interest as "cabbages." The lamentable lack of curiosity he observes in students was a prime factor in our decision to homeschool our sons.

But let me get back to my visit to East Pictou Rural High. Although spoken response from the Grade 11 students was dismal, most of the teens actively participated in the writing exercises. A good number were even brave enough to read their work aloud, which I know can be a nerve-rattling and humbling experience. The reason I'm recording this story here is that one of the creative exercises led to a life-changing moment for a student, and to a life revelation for me.

For that exercise, using a prompt from Natalie Goldberg's *Writing Down the Bones*, I had the students write the words "I like" at the top of their pages. I told them that when I said "Go!" they were to think *I like* and write the first thing that came into their minds, then keep on writing. No stopping to think about where their words should go. No editing. For seven minutes they were to ride the tide of their thoughts, to go wherever their stream of consciousness writing took them. It didn't matter what they wrote, whether it was happy, sad, confused, angry. The important thing was the exploration, the delving deeper into themselves to discover more of who they were.

A few minutes into the assignment, one girl put down her pencil. I urged her to keep writing, to which she responded that she was finished. In my firmest-but-kindest writing-instructor voice, I told her, "You're not finished until seven minutes are up. Keep writing."

The student threw me a mutinous look. Then she stared at her paper, her shoulders rigid. Finally, as other students around her wrote furiously, she picked up her pencil and began to write again. The intensity with which she wrote mounted as words flowed onto her page, until the timer I held in my hand sounded the end of seven minutes.

I allowed the students a few seconds to finish sentences and then asked for volunteers to read their work. After several obliged, the reluctant writer raised her hand, taking me by surprise. She stood and read of how she liked to sit beside a river, to watch the water, the leaves on the trees, to feel the sun on her face. The topic was peaceful, lovely. Then she read, "I like to think about my wedding day, when I'll be standing beside the man I love." I was not expecting that. Nor were the other students. The entire classroom went still. My breath caught in my throat, and I think perhaps the same was true for every student listening to a girl with untidy blonde hair speaking her dream. The very air of the room seemed frozen in profound expectancy as she continued reading about her stepfather walking her down the aisle, about her nephew bearing the rings, and about her mother crying "because she always said she knew I'd be the one to make something of myself."

I could only conclude that the switch in this student's writing focus had occurred after I'd instructed her to write more. Would she have had her light-bulb moment if she hadn't been willing to go one step further, believing, without knowing, that something of value would come from her words? During the school's lunch break, this student found me chatting with a teacher in a different classroom. After requesting permission from the teacher to speak with me, the girl, her head high, approached me. "I want to thank you for opening my eyes, and welcome to East Pictou."

As the sun's first rays lit my yard this morning, I had no inkling that a seven-minute writing exercise would change a girl's life. And I had no inkling that her courageous words would embed themselves in my mind the way they have. I may never forget her. Her bravery, shown in her willingness to share her most cherished thoughts, makes me see my *lack* of bravery as a writer, my anxiety about what others will think or say about my work. It makes me ask if *I* am willing to go one step further, believing, without knowing, that something of value will come from my words. We all walk life's path. This morning, a girl discovered she was capable of more than she knew. She discovered she *is* more than she knew. As I walk life's path through field and forest, I pray the same will be true for me.

May 28 (-3°C, clear, calm, 6 a.m.)

It's too long since I last ran. This morning I glory in swift movement, my sneakers touching down lightly on gravel as I run the back road. In the forest, I hear exotic voices of newly returned migrant songbirds—the dreamy lisp of a black-throated green warbler, the up-and-down broken phrases of a red-eyed vireo, the high-pitched wispy melody of a blackburnian warbler, that gorgeous flame-throat.

Although a heavy frost overnight scribbled hoary parallel wiggles on furrows in my garden, which Vilis tilled last evening, I see no frost on the forest floor. Sheltered beneath the countless umbrellas of foliaged branches that comprise the forest canopy, this lowest stratum of the forest exists within a more moderate microclimate than my garden and the open meadows. Our land's greatest bounty is its variety of habitats: brook, pond, grassy meadows, parkland, alder thickets, dense stands of young spruces, hardwood groves, and mature mixed forest, with all the richness of plant and animal life they accommodate. And of course, my family borrows the additional wealth of this backroad forest, enjoying its huge hemlocks and their associated community of life each time we pass through it.

When I reach Balmoral Road, sunshine spills over me. I pause a moment to listen to a white-throated sparrow's clear whistled song before running home. Fifty metres from the house, I smell my garden's freshly turned soil. Its scent is dark, thick, and filled with secrets.

In early afternoon, after finishing homeschooling for the day, I again trace the back road, this time with my sons. Dainis and Jānis scamper this way and that, their eyes keen to spot plastic, glass, and metal, which they deposit in two plastic grocery bags I carry. The metal bits likely fell off vehicles, and the other litter might have been blown by the wind or cast into the forest alongside the road by visitors, the Cathedral's owner, or loggers working in the thinned forest beyond the Cathedral. We're collecting these odds and ends so our world will be a little cleaner, a little freer from human impact.

This year marks my family's fifth of spring "junk walks." Over the years, we've gathered bag after bag of cans, beer bottles, plastic bags, and miscellaneous household waste discarded by previous owners of our property. We've also collected wind-blown paper and plastic from our yard and meadows, as well as litter from alongside our two access roads. Trash tossed by motorists driving Matheson Brook Road, which parallels our front road, often ends up on the steep slope above our access road or between it and Matheson Brook.

Of our different junk walks, it's the front-road collections that garner the most bizarre or surprising, and sometimes gruesome, items. One year, we came across a grocery bag with a store-bought chicken in it, down by the brook. Another year, we found a burlap sack with rabbit bones in it. Another junk walk yielded plastic bags filled with dirty cat litter. Two years ago, four-year-old Jāni found bags of potato chips poking out from leaves at the base of a spruce tree. "Are they garbage?" he asked, eyes wide. Since Vilis had hidden the chips in the leaves as a treat to celebrate finishing our junk walks for the year, I assured Jāni that the chips weren't garbage. We all washed our hands in the brook and then munched our treat on what we christened the "potato-chip-eating island," a gravel bar that's also handy for building rock dams and brook castles.

Now, as I watch my sons collect litter alongside the back road, it's clear that the future of our planet lies with children. Hence, the family junk walks. In addition, through my writing, I hope to instill this love for the earth in other children. When I'm leading writing workshops in elementary school classes, I read aloud *Ranger Rick* stories I've written about lynxes or vampire bats going hunting, share some of my experiences as a nature writer, and work with the children to write brief fun stories about animals. I'm always struck by the students' enthusiasm for wild creatures. When I ask which wild animal they would write about if they could choose any in the world, hands shoot up and excited or shy voices tell me cheetah, snow leopard, meerkat, frog, lion, giraffe, dolphin, moose, turtle, crab, cobra, black jaguar, platypus, and countless others I can't recall. I tell stories of when I watched a polar bear hunt seals in the High Arctic, and when I ran my hands over a tranquilized lynx's soft fur. The children delight in the tales, their eyes shining.

After my sons and I return to our yard, Dainis and Jānis dash to inspect tulips while I add our collected trash to bags of outgoing garbage. "The red one's open!" Dainis shouts. Lush and wide, the tulip steals my breath—a flamenco dancer flaunting her beauty in a slinky red dress, her black heart surrounded by yellow flames.

After praising the tulip, I at last touch my garden. Its soil is cold. Nonetheless, I plant Norland, Russet Burbank, and Island Sunshine potatoes, Stuttgarter onion sets, and Detroit Dark Red beets. This planting is a beginning. More than that, it's a connection with women, men, and children around the globe who plants seeds, bulbs, corms, and tubers in the tradition of earth-communion and sustainability known as gardening.

Every gardener knows that food doesn't come from a grocery store. Food comes from the soil. We eat plants that grow in soil; we eat animals that eat plants that grow in soil. One could say that, in essence, *we* come from the soil. The biblical verse "...for dust you are and to dust you will return"[12] flits through my mind, sparking the thought that one day red tulips like flamenco dancers might rise from my ashes.

June 2 (8°C, clear, NW breeze, 9:20 a.m.)

So many shades of green surround me—the neon lime of new aspen and birch leaves, the olive of alder and apple leaves, the rich emerald of lawn grass, and the deep, shadowed green of spruce trees, so dark in places it's almost black. Green is the colour of leading the soul from bleakness into warmth, and today my soul basks in that warmth. Vilis is teaching the boys, and I have the entire morning at my disposal—hours to explore, "look small," and allow the earth's alluring complexity to seep into my senses and thoughts. In her poem "The Summer Day," Mary Oliver wrote that she didn't know how to pray, but she knew how to observe. She knew how to touch the earth and wander in the fields, and she was blessed in doing so.[13] So am I. Each day the fingers of my life tap a new rhythm on a drumhead of family and forest, meadow and meandering, a drumhead that bounces my life-song further into this bewitching world and brings its echo back to me.

Ahead, a northern flicker flies over the back road, its golden-shafted feathers flashing on the upbeat. Voices of songbirds—robin, white-throated sparrow, American goldfinch—sound easy-going now, not harried like they did in the pre-dawn darkness of 5 a.m. A young pincherry tree in blossom beside the road is all white lace and rust-green leaves on slender stems. I walk a few steps farther, catch the scent of the pincherry's blossoms on the wind, and for an instant feel an urge to seek out the perfume's source, to rub against its blossoms and steal some of their nectar and pollen. This is the first time ever I've wanted to be a bee. I think it would be a heady experience.

I stroll into parkland beside the back road, where wild strawberries growing among spruce trees open their flowers to the sun's warmth. A tamarack sapling has clusters of new needles so tender they brush my cheek like formless rubber fingers. Grasses, goldenrod, and *Maianthemum*—wild lily-of-the-valley—grow among cushions of red, gold, and green mosses. The *Maianthemum*, which usually grows in a forest's understory, surprises me until I recall that our parkland is on its way to becoming a forest.

On disturbed land, the change in vegetation over time is called succession. This change continues until a stable plant community, such as a mature forest, establishes and maintains itself. A stable, end-of-succession plant community is known as a climax community. Our parkland is a good example of old field succession, where what was once a farm field, as evidenced by overgrown but still discernable plow furrows, has been invaded by grasses, wildflowers, shrubs, and trees. Someday what is now parkland will resemble the woodland alongside the back road, where the climax community is coniferous forest, although stands of mixed wood also occur. In a few places on our property, hardwoods create their own small, localized climax communities, such as the sugar maples in the Amphitheatre.

Each climax community supports its own characteristic web of organisms, its own microcosm of life. When birding in mature hemlock forest like that along the back road, I'm likely to see or hear black-throated green warblers, golden-crowned kinglets, ovenbirds, and eastern wood-pewees. On the other hand, when I'm at the Amphitheatre, I'm likely to jot in my notebook sightings of American redstarts and spunky wood-warblers called northern parulas, whose territorial song is a rapid buzz that ziiiiiips upward in pitch and abruptly falls off.

In successional plant communities, such as our parkland, plants provide habitat for bird associations that change as the vegetation changes. For example, an alder thicket on a former field behind our house now provides habitat for shrub-loving chestnut-sided warblers; however, spruce saplings growing among the alders will one day overtake them and diminish the habitat quality for chestnut-sided warblers. And in the parkland south of our house, where I'm walking, the voices of song sparrows and American goldfinches brighten this morning, but those species will not likely be present in the heart of mature coniferous forest that will one day cover this land unless Vilis and I intervene.

I sit on the ground in the sunshine and watch ants and a spider explore grasses and moss carpets. This is the ants' world, their stratum, their layer of the earth. As an undergraduate zoology student, I was fascinated by the findings of an ecologist named Robert MacArthur, who investigated whether five warbler species could coexist without competing for insects in their woodland habitat. (Perhaps even then I was subconsciously a birder.) Intriguingly, MacArthur observed that Cape May warblers foraged close to the tops of spruce, fir, and pine trees, whereas yellow-rumped warblers more often foraged low in the trees. Blackburnian warblers hunted insects on the trees' outer branches, while bay-breasted warblers foraged for food on the trees' interior branches. Black-throated green warblers tended to hunt in the trees' mid-section, both vertically and in terms of distance from the trunk.[14] Each of the five warbler species preferred to hunt insects in a different part of the trees; thus, each occupied its own niche, an ecological term meaning the functional role of an organism within a biological community. What you and I might see simply as a tree, the warblers saw as a collection of distinct hunting areas, enabling the birds to coexist and avoid competition for food.

A thought born of MacArthur's study jolts my imagination. Perhaps life is a tree, and we're warblers occupying different niches in time rather than space—sometimes riding high at the top of the tree, and sometimes skidding down to the bottom; sometimes hugging the safety of the trunk, and sometimes walking out on the edge. At times in my life, I've been bold, and at others, timid. At times I've celebrated successes; at others, I've suffered setbacks. Throughout our time on earth, the tree will continue to offer its branches. Perhaps we need only choose which warbler to be.

I shift my body to a more comfortable position in the grass and notice a large anthill a metre away. Our property has many such anthills. This one is about thirty centimetres high, is tufted with grass, and has a goldenrod stalk growing out of a bald spot of brown soil. On at least two occasions, I've flushed ruffed grouse from anthills similar to this one and wondered if the birds were anting, a form of pest control used by some bird species. Robins and blue jays are known to catch ants who, in defence of being caught, squirt out chemicals such as formic acid. The birds rub these chemicals onto their feathers to rid themselves of lice and mites.[15] What first prompted a bird to try this behaviour? Did a robin observe a blue jay anting and then try it?

Nature is a vast realm for experimentation by its denizens. Michael Gratson, the sharp-tailed grouse researcher I assisted in Manitoba's quiet farming country, once commented to a local newspaper reporter, "There's more happening out here than in Winnipeg."[16] If only we have the eyes to see and the ears to hear.

Here's an example of new learning by a wild animal that occurred right in my own yard. Initially, the leaves of my grapevines were unmolested by wildlife. Then, one year, one or more white-tailed deer sampled the grape leaves and developed a taste for them. Now, each summer, deer feast on my grapes unless I fence the herbivores out. The natural world isn't static. Plants and animals respond to their environment. Goshawks typically hunt grouse and pigeons, but they also take our chickens. Deer that never used to eat grape leaves in my yard now seek them out. For our wild neighbours, both survival and life quality hinge on animals being closely attuned to their immediate surroundings and being able to pivot to take advantage of new possibilities.

An intriguing way in which animals pivot to take advantage of new possibilities, especially foods, is through the formation of search images, a term coined by Lucas Tinbergen. In 1960 Tinbergen reported observing European tits eating certain forest insects and ignoring other available insects. He proposed that the small songbirds, after encountering a locally abundant insect species randomly many times, cued into certain features of the insect's appearance and then subsequently focused exclusively on those features. This behaviour enabled the tits to detect that insect quickly and efficiently.[17] I think of my own birding experiences, during which I look for specific field marks, such as the pattern and colour of markings on a bird's head, to speed my identification. The concept of a visual search image makes perfect sense.

But what about smell? Do animals also develop search images for the scent of a particular food? This was the question underlying Vilis's doctoral research. As part of his studies of the hunting habits of striped skunks, he tested whether skunks form olfactory search images, that is, whether they cue into the smell of a locally abundant food and then focus on that scent, subsequently developing the ability to quickly detect that food at increasingly greater distances. His research partners for this exploration into uncharted animal behaviour territory were Emmy Lou, Izzy, and Tristan, three semi-tame skunks Vilis descented when they were kits. The *Mephitis mephitis* trio tolerated having lightweight string leashes attached to their collars. This enabled Vilis to carefully watch their responses to their environment *and* prevent the skittish skunks from panicking and racing off whenever they heard the roar of a vehicle nearby.

Raised in kennels in the back porch of an old farmhouse we were renting at the time, the skunks were familiar with dog kibble as food. During the olfactory search image experiments, Vilis led one hungry skunk at a time from the back porch onto the farmhouse lawn and observed his research subject's foraging behaviour. The skunks were ideal animals to work with for this experiment because they always gave a clear indication of when they scented the experimental food. We repeatedly saw them do the following: Tristan, Izzy, or Emmy Lou snuffled about in the grass, nose to the ground, but as soon as the skunk detected the scent of food in the air, its head came up and it moved in a direct line toward the food. For the experiment, Vilis first tossed pieces of kibble right in front of a foraging skunk, allowing it to find the food. On subsequent test days with each individual skunk, he tossed kibble farther in front of it and observed that the skunks learned to detect the scent of kibble at increasingly greater distances; that is, they formed an olfactory search image. Through other experiments, Vilis also showed that when exposed to a new locally abundant experimental food, such as bits of bologna, the skunks lost their olfactory search image for the kibble and formed a new one for the bologna.

Three decades after Tinbergen reported visual search images in European tits, Vilis's research reported the first instance of an animal forming an olfactory search image. As exciting as this discovery was for Vilis, the adaptive behaviour shown by the skunks was, for them, simply part of life. Heads down, noses snuffling loudly in the grass until the irresistible smell of kibble or bologna reached their nostrils and caused them to lift their heads like an English pointer and reveal the distance at which they detected their "prey," Izzy, Emmy Lou, and Tristan were simply pivoting in their foraging behaviour as they reacted to food availability in their environment. In doing so, they happened to provide a keen young zoologist with an illuminated glimpse into the life story of the striped skunk.

While I sit near the anthill thinking about search images, I hear the rollicking song of a male ruby-crowned kinglet singing in a stand of tall evergreens on the eastern edge of the parkland. The ruby-crown has a big voice for such a small bird. The size of a mouse, this kinglet species, with its olive plumage, white wing bars, and prominent white eye ring, belts out its song, usually from high in trees. The first time I identified a ruby-crown I spent a half-hour creeping through a forest in the Yukon, gawking through binoculars at treetops, before I traced the song to its diminutive, flitting singer. "So, it's you!" I was expecting something larger and more aggressive, something more imposing and impressive.

How many other times have I underestimated a singer, a speaker, or a runner when their notes, their words, or their feet were still? I like to think that my children will surprise me with unexpected talents, and that maybe I'll surprise them, too.

I stroll back to the yard, where Vilis is timing Dainis and Jānis while they pedal fifty-metre bicycle dashes, first from a standing start, then from a moving start, and finally from a moving start when wearing snug sneakers rather than loose rubber boots. Grins light my sons' faces. Dandelions speckle the lawn. Mares' tails streak the sky.

June 8 (-1°C, clear, calm, heavy frost, 5:30 a.m.)

A killer came in the night, masked, and cloaked in darkness. The chickens alerted me to its attack at 2 a.m., their frantic squawking reaching my ears through an open bedroom window. I shook Vilis's shoulder. "There's something in the coop!"

Vilis rushed downstairs and ran to the coop with rifle and flashlight in hand. I heard a shot followed by agonized screams, then two more shots followed by silence. The bobbing flashlight returned to the house, and Vilis climbed into bed. "A raccoon got Young Samson."

Heart pounding, I asked, "Who did the screaming, the raccoon or Young Samson?"

"Young Samson. The coon crawled in between the wall and that heavy wire that screens the doors. I fixed the gap." We'd left the coop's top doors open for fresh air.

In a strange way, it seems that even in death there can be order and perhaps the granting of a subconscious wish. Of the nine chickens in the coop, the ring-tailed, black-masked raccoon chose Young Samson, whom Vilis and I had decided to butcher to restore harmony to our flock.

Now, after the attack, we try to sleep but can't, our minds filled with the night's violence. When the alarm rings at 5:30, we're still awake. Beyond the bedroom windows, mist hangs in the brook valley, and frost whitens the lawn and meadows.

We rouse Dainis and Jānis from sleep, and together with them eat a quick hot porridge breakfast. Then Vilis and I load his aluminum Grumman canoe onto our pickup. Eerie loon calls echo in the cold air as we put the canoe in on calm mist-covered Earltown Lake, a fifteen-minute drive south of Ravenhill. We paddle through the mist and spot the loons' sleek dark bodies diving and then reappearing farther away on the lake surface. We drag the canoe a short portage to the lake's second belly and paddle until we've explored its reaches, until sunshine has licked away the mist, until we must return home. At 10 a.m., the boys are on the soccer field in Pictou. At noon, we're home again, eating "magic sandwiches"—chicken, mayonnaise, lettuce. Two hours later, the four of us are cycling the abandoned railway trail west of Tatamagouche.

Some days, life is so full it spills over all around us. On those days we're caught up in the intensity of existence, perhaps spurred on by recent trauma or death. If I hadn't heard Young Samson's screams in the night and the gunshots that ended his agony, would dawn have seemed so welcoming and this day so precious? Without that chilling reminder of mortality that kept me awake for hours, would I have suggested this cycling excursion after already canoeing on Earltown Lake and attending the boys' soccer matches in Pictou? Likely not. Death shocked me into a richer appreciation for life, into embracing all this day can offer. Mary Oliver ended her poem "The Summer Day" with the line, "Tell me, what is it you plan to do with your one wild and precious life?"

Cherish every moment. Ahead of me, Jānis powers his bicycle forward on the trail, his short legs pumping up and down like living pistons as he leads me further into my "one wild and precious life."

June 11 (ca. 10°C, clear, 7 a.m.)

This morning I have company as I walk the back road. Two deer bound slowly from the south meadow into the woods at my passing, and songbirds proclaim their possession of breeding territories. Pollen lies in lacy yellow froth at the edges of now-dry puddles on the back road, puddles that formed after a shower washed smoke from the sky early this morning. The smoke drifted from Quebec, a thousand kilometres away, where already this year 31,000 hectares of forest have burned, five times the average area burned over the past five years, according to CBC Radio. The smoke blanketed our hills, closed in the sky, and awakened me with headaches. Thanks to the early morning rain, our air is now clean and fresh.

Even cooler air invites me into the shady forest, with its towering hemlocks. I round a gentle curve and stop short on catching a glimpse of red and blue machinery. It's so rare that anyone other than my family uses this road. When someone else does, it's nearly always about logging. I walk closer.

Seated high above the back end of a blue logging truck, a man is working a red log grapple—a pair of massive metal claws. Using the grapple, he lifts two, three, or even four recently cut hemlock trunks from the side of the road in the same way that a finger and thumb lift toothpicks. He swings the grapple over to the truck and places the logs on the truck bed, nudging them this way and that, positioning the load to receive more logs. He doesn't see me and continues his task while I watch, impressed with his skill. Then he shifts his seat more in my direction, and the action stops. He's seen me. After a moment, he resumes his work, and I walk home.

I tell Vilis about the grapple man, and my husband suggests that I take our boys to watch him. So, Dainis, Jānis, and I mount our bicycles and pedal fast down the back road, only to find we're too late. No longer seated high above the truck, the white-haired grapple man stands beside the vehicle, his round face smiling. He motions us over.

"You're pretty good with that," I say, pointing to the grapple. "I thought my boys would enjoy seeing you work, but it looks like you're done."

He smiles modestly, shrugs, then points to the other side of the road where a few thin logs remain. "Still got a few scaffolds to pick up. You got the old Noble Hayman place?"

"Yes." I brush a black fly off my arm, the miniscule biting pest, one of hundreds gathering around us as we stand motionless in the forest.

"Thought so. I used to know some of the old ones around here." The grapple man pulls two thick straps from beneath the truck and stretches them out on the ground. Then he climbs the ladder back up to the seat where I first saw him perched atop the truck's back end.

The boys and I back our bicycles away. Black flies swarm around us as we watch the log-loading expert pick up the scaffolds, one or two at a time, and carefully fit them into spaces in the load. While we fidget and stomp to nudge flies away and wave at the black clouds in front of our faces, the grapple man calmly plucks, in turn, each of the two straps from the ground, drops it on the load, and positions it carefully on one side of the load. He then swings the grapple to the centre of the load and drops it, causing the whole truck to shake. After placing the wide straps over the load and grapple, he fastens the straps, climbs down the ladder, waves to us, and enters the truck cab. With the action over and the black flies unbearable, my sons and I pedal furiously homeward.

A snake exists in every Eden. At Ravenhill that snake is black flies. We first viewed what is now our property in mid-August, when fly season is usually over. We took possession in early September and encountered wasps, but not biting flies. The following May, black flies besieged us, crawling into our ears and leaving huge welts on the boys' necks. We fought back with hats and hoods, long sleeves and pants, head nets and aromatic Skin So Soft moisturizing oil. We learned that the cool hours of evening and early morning were the best times to venture outdoors with bare limbs, and I found a healing salve to take away the bites' incessant itchiness. We prayed for strong winds to ground the little blood drinkers and mowed back the grass in which they hid. Still, all too often, the boys and I ran for the house, with Jāni squealing and Daini and I yelling, "I *hate* these flies!"

We still yell sometimes, and run. And every spring, Vilis and I talk about screening in part of the porch. Even so, we know that black flies are an important component of the natural food web. Their larvae are eaten by fish and aquatic invertebrates, and the adult flies are consumed by swallows and other insectivorous birds. This insect that's a pest to my family is a boon to some of Ravenhill's wildlife species. Right now, though, I'm not thinking about fish or swallows. As my sons and I race through the forest on our bicycles, my only thought is to outrun the black flies seeking my blood.

June 12 (6°C, clear, calm, 7 a.m.)

The backroad woods are a cool well of greenness that I drink of deeply, swim through slowly, and pirouette within, lifting my arms high in celebration of a new day and this world. Before I entered the woods, a hare hopped from the meadow as I passed, its white feet flashing. And here in the forest, songbirds serenade, their melodies echoing in this woodland well in a way that avian voices never do in the open parkland.

This is a season of "romance." Yellow pollen lies at my feet, cast down from conifers. White pincherry petals are confetti on the stones over which I walk. Birds—black-throated green warbler, ovenbird, ruby-crowned kinglet—fill the air with notes of challenge and desire. A wild tide of reproductive energy surges through the morning, its aim to pass on genes to future generations of conifers, pincherries, and songbirds. I pass through the tide's swelling wake, a silent witness to life's propagative fire.

It's 10 a.m.—celebration time! "Ravenhill Academy" is closed for the summer. The boys and I sit on the porch, with the green and white lace of spring before us and the cool sweet thickness of ice cream in our mouths. After we finish our treat, I plant herbs and flowers while Dainis and Jānis romp and chatter around me.

Vilis and I decided to homeschool Dainis when he turned five. For me, that decision was largely fuelled by our older son's intense shyness. For Vilis, it was to prevent Dainis from becoming a "cabbage." After the first year, I continued, tailoring Dainis's and eventually Jānis's learning programs to reflect their individual strengths and weaknesses. Vilis and I want our boys to have time to play, to be children, to get to know the natural world, and for the four of us as a family to know each other well. Homeschooling hasn't been all joy and success. Education involves the tension of going from not knowing to knowing, and the path to illumination can be stubbornly tortuous. Sometimes I wonder if we're doing the right thing or if we're doing enough, but at other times I know that Vilis and I made the right choice for our family and that our sons are learning things we never could have imagined.

Rachel Carson, a luminary of environmental education, wrote, "If a child is to keep alive his inborn sense of wonder without any such gift from the fairies, he needs the companionship of at least one adult who can share it, rediscovering with him the joy, excitement and mystery of the world we live in."[18] Vilis and I are two such adults, intent on immersing our sons in the excitement and mysteries of nature. For this, our meadows and forest, brook and pond offer a variety of classrooms wherein learning includes camping under the stars, collecting maple sap, saving a rooster from a goshawk, and observing drop prisms like bits of rainbow hung in the forest.

Hours later, in early evening, big scattered raindrops catch Jānis and me near the bridge after we collected mail. Unfazed by the rain, Jānis natters brightly, his sturdy legs skipping, his black T-shirt and blotchy blue shorts vivid colours against the greens and browns of the front road. The raindrops fall heavily onto a pincherry beside the bridge, sending its white petals fluttering down—like love—all around us.

There are times when I rail against the bonds of motherhood and teaching, when my spirit cries out that I've become so lost in my children that I no longer know who I am. During those times, I yearn to escape domesticity and seek wilderness instead, with its new sights and challenges. Two months ago, when I began this journal, was such a time. And yet, there are many times, such as now on this walk with Jānis, when simply being with my sons affords nearly overwhelming joy. I always knew I wanted children, yet I had no foreshadowing of the consuming emotion that would flare to life inside me when I looked into my firstborn's eyes and held him to my breast. I waited seven years after marriage for a child, and Dainis's arrival cracked open a new sphere in my universe, allowing in the brilliant flame of mother love. And dear Jānis, born two-and-a-half weeks early and only two days after his brother came home from ICU after suffering a potentially life-threatening illness, gurgled sweet baby songs as he gazed into my face. The next time I rail against motherhood and teaching, I hope I remember those songs.

June 13 (12°C, overcast, S wind, 8:30 a.m.)

A damp wind tells of rain on the way. Do the birds sense it? Their voices sound rushed, harried, crammed together into space and time too small to hold such a variety of sounds—trills, buzzes, whistles, flute-like melodies, and even the winnowing of a snipe's wings. Before Vilis and I bought our land, I listed features of what I thought would be the perfect property. The list included a long view, solitude, a river or lake, good soil, and a mix of habitats for songbirds, particularly brightly coloured wood-warblers. Here at Ravenhill, we found every item on the list.

Today, I have another entire morning to myself, another opportunity to explore our land at my leisure and discover what the world will show me and teach me. Apple trees, both wild and domestic, are in bloom, their flowers edging the woods and yard with frilly pinkish-white lace. I walk among them to the edge of the young spruce forest overtaking alder shrubs west of our house. In a meadow adjoining the forest, new goldenrod stalks reach my knees. At the forest edge, alder leaves, like those I wished to shove into my mouth three weeks ago, are five centimetres long, with more unfurling. Spruce trees bear new shoots the colour of green after-dinner mints, their tips wearing pointed bronze caps of protective bud scales not yet shed. Spring, with its burgeoning, flourishing new growth, will soon give way to summer and all its riches.

At the forest edge, a male Tennessee warbler with grey head and olive back snaps out its three-part "sweet-sweet-sweet, chip-chip-chip, chirrrrrr" song. Like the ruby-crowned kinglet and ovenbird, the Tennessee warbler is a small bird with a big voice. A trail cut by Vilis and the boys last winter lures me into the forest and leads me alongside a bank of moss-covered rocks that edges one of the old fields on our property. Who dug these rocks from the land and loaded them onto a stone boat or wagon and hauled them here? This property has known many before my family. What did it look like when Scottish settlers claimed it?

The man who built our house, John Hayman, was one of twelve children born to William and Margaret Elizabeth (nee Maillard) Hayman. Hailing from Argyllshire, Scotland, William Hayman left his homeland and arrived in America around 1777, where he served as a British loyalist during the American Revolution. After the war ended, he immigrated to Canada, arriving in Halifax in 1783. About two years later, he moved to the Tatamagouche area, where he married Margaret Maillard, farmed on rented land, sold trees for ship building, and petitioned for a land grant. His son, John, was born in 1819, married Mary Slade in 1840, and raised five children, one of whom was George, the eventual father of Noble Hayman, mentioned by the grapple man as one of the "old ones" he'd known around here. Noble married Ellen Fraser, and it was this Mrs. Ellen Hayman who told Vilis and me that it was John

Hayman, her husband's grandfather, who around 1841 built the main part of what is now our house.

Like many other Scots who settled in northern Nova Scotia, the Haymans were farmers. They raised livestock and grew grain on land that looks out over hills, and from which it's easy to imagine a piper playing reels and laments. In 1969, the property was sold out of the family, setting in motion a chain of purchases and sales that eventually led to my family buying it twenty-two years later.

Petite and quiet, Ellen Hayman visited Ravenhill several times with her daughters just to see the old place, after my family took possession. She shared snippets of Hayman family history and allowed Vilis and me to copy old photographs of the house. A picture dated 1945 shows an elegant farm home graced with a wrap-around balcony and matching porch. In three later photographs, the upper balcony is gone—torn off by a hurricane. At some point, the wrap-around porch also disappeared. Story has it that the then-owner tore off the porch to reduce his taxes. A photograph from the 1970s shows a modest-looking home, with a tall wire pen that appears to be a hen yard attached to the front of the house. Apparently, the owner at that time housed chickens in what is now our kitchen.

A flurry of sound and movement ahead of me halts my musings about our property's history. Hissing loudly, a ruffed grouse erupts into the air, wings fluttering awkwardly. Ah, this is a hen whose broken-wing display is a ploy to lead me away from her chicks hidden somewhere nearby. On other occasions, I've spotted young grouse chicks—tiny balls of downy feathers frozen against the ground, half-hidden in dead leaves or under fallen branches—but I don't see any of this hen's brood. Still hissing and fluttering, she disappears into the alders. I continue my outing, stepping carefully.

At Ravenhill's western edge, I turn north and follow a boundary of crumbling fence boards. The fence separates a neighbour's old-growth hemlock forest from a stand of spruce saplings that tells the story of yet another pasture or grain field on our property reverting to forest. Our land is a patchwork of such fields bordered by woodland.

With early morning's mosquitoes spurring me on, I thrash through the dense saplings, their branches whipping my face, then link up with the top of the old wood road down which we pull our toboggan on the way to winter camping or collecting maple sap. The Amphitheatre stretches before me, its curved wall enclosing a soaring green span of hardwood canopy. When I reach the Amphitheatre, instead of sliding down its steep wall, I turn west and walk a path strewn with pincherry and wild apple petals, past the campsite we used for our spring campout, and onward as far as the beaver dam that spans Matheson Brook downstream from our sugar maples.

With pole in hand, I step onto the dam and slowly make my way along it. An apple petal floats at my feet, caught in still water at the edge of the broad pond pooled behind the dam. The dam itself is an engineering marvel—a fifty-metre barricade of mud, stones, chewed branches, and leaves stacked nearly two metres high. A fallen tree that used to lie at the brook's edge is now an island in a sea of beaver water. Before the dam obstructed the brook's flow, my family brought picnic lunches to the fallen maple, clambered among its spreading limbs, and lay back on thick branches while soaking up sun and brook air. Now we would have to swim to do that.

When I reach the end of the dam, I step into a tangle of chokecherry saplings and pause to watch an American black duck come in for a landing on the beaver pond, water skimming in spray off its reaching feet. The duck's low, gurgling quacks mix with the "witchy-witchity" song of a male common yellowthroat—an olive-yellow wood-warbler with a black mask—that's skulking in the chokecherries. Water swirls to my right—a beaver submerging—and I notice a pile of branches heaped onto the brook's bank. That's the beaver lodge. Perhaps, right now, beaver kits are sleeping in that lodge. Ripples upstream alert me to the presence of another beaver, so I stand perfectly still as it swims toward me, a V-wake rippling behind it. In a sudden, silent move, the beaver dives below the water's surface, and I don't see it reappear. Perhaps it caught my scent.

The black duck on the pond growls its nasal quack. A fish jumps and splashes back into the pool. Perched in a tree beside the beaver pond, a purple finch tosses its bright melody into the air. These sights and sounds, like the apple blossoms, Tennessee warbler, and grouse hen, bring me a sense of belonging, a sense of community with the land on which I live. The riches of nature are my riches, refining my soul and sparking light into my spirit. I know I'm a nanoscopic brush-stroke in the enormous painting of life on earth, but if I didn't have this bit of nature here on my doorstep—the birds and beavers and towering hemlocks—I would be even smaller.

People have long been drawn to nature to replenish their spirits and figuratively fill their wells and sharpen their saws. The psychologist and psychoanalyst Erich Fromm first used the term *biophilia*, defining it as "the passionate love of life and of all that is alive."[19] Edward O. Wilson, an evolutionary biologist who studied the social lives of insects, particularly ants, popularized the term in his book *Biophilia: The Human Bond with Other Species.* Biophilia describes how people are instinctively drawn to other living things, how a connection with the rest of nature fulfills a basic emotional need within us.[20] Perhaps that emotional need is what I sought to fulfill on the day I began writing this journal. At that time, when I set out on my walk along the old wood road beside Matheson Brook, a deep craving tore at me, yet when I returned home that day, creative energy and a sense of unexpected fulfilment surged through me. The escape from homeschooling and parental responsibilities showed me the power of nature to nourish, heal, and show me more of who I am.

Now, the morning waxes on, so I head upstream from the chokecherry thicket by the beaver pond. Logs and branches lie stranded on dry ground, carried here by the brook at flood time. The waterway dwindles in breadth and depth until—at a sharp curve in the brook I call the Shepherd's Crook—it's shallow, no more than three metres wide, and split in two. I cross easily on stones and logs, then climb the steep rocky slope below the old wood road. How many of these rocks were originally here, and how many were thrown down onto the slope by hardened hands clearing fields for crops? Many are the latter, I suspect, for the ground on the far side of the old road, at the top of the steep incline, is not nearly so rocky. From the wood road, I gaze back down to the brook. Trees below me entice me to linger, but my homeward steps are quick and impatient now. It's time to draw.

Life sometimes provides us with opportunities to don a different coat, take on a new character, or occupy a different niche. My new coat is black and white, the signature colours of the pen-and-ink artist. It rests on my shoulders like the borrowed cape of an exotic neighbour. Last year I received an unexpected invitation to present an exhibit at the 1889 Gallery in Tatamagouche in early August of this year. Previously, I'd only ever hastily prepared a few drawings for the gallery's Members Show, usually putting my pen to paper the night before each show's opening. Now I have the gift of time but still have several pieces to complete, since I agreed to present at least sixteen.

Nearly all my drawings depict natural subjects, and in a way, each illustrates both my fascination with nature and tells a chapter of my life story. The first new drawing I completed for the show depicts a great horned owlet I photographed in the Yukon, where Vilis and I lived in a log cabin on the shore of Kluane Lake. Another drawing illustrates a red squirrel swimming in Nova Scotia's Kejimkujik Lake, a squirrel Vilis and I paddled alongside during our honeymoon, nearly two decades ago. Still another work illustrates a killdeer's nest beside a shallow lake near Hay River in the Northwest Territories, where Vilis photographed it during the time that he and I worked at the now-defunct Heart Lake Biological Station and fell in love. Another drawing shows one of the red fox pups we cared for and released in southwestern Manitoba, and another depicts long-legged crane flies that Vilis photographed when we lived in Victoria. Today I'll draw lichens growing on old wood, a "look small" subject that Vilis photographed when we lived in that flat beside Chocolate Lake, outside Halifax, where a junco's monotonous trill catalyzed me into birding.

I don't think of myself as an artist, but I crave beauty and yearn to create beauty. So, today I'll position dots of black ink on white paper, one at a time, hoping, as I always do, to capture some of nature's elegance and drama. More dots for darker areas of wood; fewer dots for pale areas of lichen. And all the while, the new coat resting on my shoulders will make me feel elegant and dramatic, like black and white.

June 17 (10°C, overcast, strong S wind, 5:15 a.m.)

The first time ever that I walked on flowers was in Victoria, where ornamental cherry trees' pink petals lay strewn over sidewalks. I stared in wonder at such cast-off beauty and stepped onto it lightly, as though treading on a piece of heaven. Yesterday, apple blossom petals drifted like pink-tinged snow into the garden while I planted carrots, and clusters of petals created delicate halos around Dainis and Jānis as they climbed in the huge old apple tree beside the garden, which is the source of the petal snow. After I finished with the carrots, more petals drifted down onto on the lawn like pale pink confetti while the boys and I strolled to the house.

Today holds none of yesterday's gentle peace. Instead, urgency surges through me. It's newly light outdoors, and I'm racing the elements—wind in my face, rain in the forecast—in a bid to finish planting my garden before the storm arrives. Did my mother feel this same sense of urgency? With nine children to feed, she planted and tended a huge vegetable garden and expected me and my siblings to help with the weeding and harvesting. Sometimes, though, with a brother or sister at my side I mischievously squirmed through the garden on my belly, raiding crisp pea pods or cucumbers. Now, I wonder if my mother ever watched us carrying out our raids but never said a word, allowing us to savour our innocuous adventures.

In my own garden, I quickly set out tomato, pepper, broccoli, cauliflower, cabbage, eggplant, ground cherry, and sweet basil bedding plants that I started indoors months ago, the trays of soil and seed soaking up sun on our living room's deep windowsills. At 7 a.m., I pause for breakfast with my family and then hastily sow peas, beans, and corn. At 9 a.m., Dainis, Jānis, and I are at Speakman's Nursery in Tatamagouche, choosing cantaloupe and watermelon starter plants. Back home, while the boys complete small garden tasks,

I seed squashes, pumpkins, cucumbers, dill, fennel, radishes, and finally, transplant more broccoli and the melons. At noon, wild wind whips trees and plasters us with warm rain, but we laugh and cheer, our arms raised high in celebration. We did it! Let summer come. The garden is in.

2

As Though It Were a Kiss

June 21 (8°C, clear, calm, 4:55 a.m.)

The moon rides high, its white face round and full on this longest day of the year. For once I've risen before Samson's crowing splits the air. I hurry outdoors to await sunrise and stand on the gravel drive, with the moon on my right and the brightening sky on my left.

Pandemonium breaks loose in the coop. I rush to the feed room and flip the light switch. Nothing. I open the coop door, expecting devastation within, but the six hens and two roosters are all perched high and safe. Then I see it—a raccoon scrambling along the beam above my head. Its eyes meet mine. Then the furry intruder squeezes through a hole in a chicken-wire barrier between the coop and feed room, scrambles along a beam, runs across the top of storage shelves, and scrabbles through the cat opening that Vilis reduced in size by half after the last visit by a masked raider. I yell, speeding the coon on its way. Then I push the chicken wire back into place and make a mental note to staple it firmly to the coop wall and to stop feeding our cats in the feed room.

The poultry pandemonium subsides. Samson flies down and begins his ritual of greeting the day. Last week, Dainis, Jānis, and I read an African folk tale about how a mouse, fly, and spider tricked the king of the sky people into handing over the box that held the secret of light. When the scheming trio opened the red box, they found a rooster inside. He crowed, and the sun rose, lighting the world.

Outside the coop, cold wind brushes my face as I scan our yard while waiting for sunrise. Apple blossoms are all but gone. Developing cherries are hard green beads. Chives and lilacs flaunt mauve blossoms that will entice hawk moths and viceroy butterflies later today, after the air warms. Samson continues to crow, and finally golden light spills over the eastern horizon, painting birch trunks a startling white gold. Minutes later, a pale golden orb clears the horizon.

Perched on the edge of Nova Scotia's hottest months, summer solstice offers the promise of long, languorous days, picnics and swimming, and the vibrant flavours of fresh fruit and vegetables. It also brings a warning that despite the heat and bounty still to come, the amount of daylight at our disposal will shrink as darkness increasingly steals away the edges of evening and morning. I quash the thought. Why think of losing summer when it's only begun?

In the west, the moon is now no more than a pale shadow of the bright face I saw earlier. Last evening, that bright face was a ripe sphere that rose while Jānis and I baked oatmeal chocolate chip cookies. (Vilis and Dainis are at Cub Camp.) It was Jāni who spotted the huge yellow ball that seemed at first to rest atop hills on the horizon before slowly rising into the darkening sky. At bedtime, after the cookies were baked and Jāni sampled one, we turned out the lights and made our way through darkness to the stairs leading to our bedrooms, our path lit by moonlight that fell in neat, straight-sided tracts onto floors and walls. Big moon, you brought no late-June frost this year, as you have in previous years.

Black flies drive me indoors. I drink hot water, sample a cookie, and stare out the kitchen patio doors at wind-whipped tree leaves and glistening lawn. The grass, with its dewdrop diamonds, provides a showy foil for gangly dandelion stems topped with fuzzy seed orbs.

My snack finished, I don a net jacket and weed a flowerbed near the foot of a pine windbreak east of the house, spading out grass clods and trowelling out dandelions, goldenrod, and yellow wood-sorrel, tossing them into a wheelbarrow. All these plants are invaders and opportunists of the most frustrating and successful kind. Like a never-ending wave of colonizers, they spread beyond the limits I've laid out for them and into the disturbed soil of my flowerbed via fast-growing shoots or wind-blown seeds. Ecologists call such species ruderals, derived from the Latin *rudera*, meaning "rubble," a reference to the disturbed ground on which such plants grow. Gardeners call them weeds.

The earth's array of plant life is so diverse, it sometimes seems an overwhelming task to make heads or tails of it. An evolutionary botanist named J. Philip Grime came up with a plan to create order out of all that diversity. Grime grouped plants into three categories based on their growth and reproduction in response to two environmental factors: disturbance and stress. He referred to the three categories as plant strategies and described them as the three points, or vertices, of a triangle. At one vertex of Grime's triangle are the ruderals or weedy species. Like the weeds in my flowerbed, ruderals are found in low stress/high disturbance environments. They grow quickly and produce many seeds. On the flip side of causing headaches for gardeners, ruderals serve an important role in establishing plant cover on severely disturbed soils like those left in the wake of natural disasters, such as volcanic eruptions and avalanches, or man-made disturbances, such as road building or house construction.

At another vertex of Grime's plant strategy triangle are the stress-tolerators. These plants grow in high stress/low disturbance environments. Stress-tolerators are long-lived, slow-growing plants that survive and reproduce under extremely harsh environmental conditions. Think of krummholz, the stunted wind-sculpted alpine trees growing at timberline in the Rockies, or tuckamore, the undersized evergreens wind-woven into tangled mats on Newfoundland coastlines.

At the third apex of Grime's triangle are the competitors. These are highly productive and often long-lived plants that establish and maintain themselves in low stress/low disturbance environments. Think of undisturbed native prairie grasses or native forest trees.

Of course, not all plants fit neatly into Grime's three basic strategies. Many species straddle the line between two strategies, possessing some characteristics of each. Some even share characteristics of all three strategies and fall in the middle of the triangle.

During the two summers I spent on Ellesmere Island, I became well-acquainted with plants that illustrated Grime's three strategies. The research site, a lowland oasis of High Arctic vegetation adjacent to Alexandra Fiord, was situated at nearly 79° North latitude and experienced long dark winters and short summers with twenty-four-hour sunlight. Temperatures were cold even in summer, even with all that sunshine. The average air temperature in June and July was 6°C, and in August, 3°C.[21] Vegetation grew close to the ground surface, taking advantage of warmer air temperatures and slower wind speeds at soil level.

You might think that any plant growing at such a high latitude would be classed as a stress-tolerator, but even at "Alex," examples of ruderals, competitors, and stress-tolerators abounded. A tiny herbaceous perennial with an astonishing number of flowers and, later, seed pods, *Draba groenlandica*, grew in gravel surrounding the buildings of the abandoned Royal Canadian Mounted Police station that served as our research base. This plant was undoubtedly a ruderal. Sedges and grasses dominated meadows on damp soil near rivers. The dense, lush growth of these perennial graminoids (grass-like plants) during the two-month growing season proclaimed them as competitors. Tiny-leaved, ground-hugging cushion plants like purple saxifrage, *Saxifraga oppositifolia*, and moss campion, *Silene acaulis*, exemplified the slow-growing, "defensive" plant strategy of the stress-tolerator. So did arctic white heather, *Cassiope tetragona*, the subject of my research. A prominent species in the lowland's vegetation, arctic white heather thrust its shoots no more than twenty centimetres into the air, and much of each shoot was covered with tiers of old leaves that aided in collecting insulating snow to protect the shoots during the long High Arctic winter.

Because I'd studied zoology, not botany, as an undergraduate student, my botanical research involved a steep learning curve. Lured into my project by the prospect of working in the High Arctic, I found myself poring over plant-identification guides, measuring heather growth in increments of millimetres, snipping plant samples to calculate how nutrients were allocated to various plant parts, and generally spending a lot of time lying on the treeless tundra, eyeballing High Arctic plants. The wide-open, nearly two-dimensional landscape reminded me of the prairies. It had the same "big sky" ambience that encouraged long views and inspired me, during breaks from measuring heather growth, to count seals basking on the fiord ice in the distance.

During my time at Alexandra Fiord, rare 10°C days allowed me to remove my coat and sweater and feel the sun on my bare arms, but generally when I was outdoors cold was my constant companion. My hands developed red itchy chilblains from repeated contact with frigid damp soil and plants. Washing my hair in a glacier-fed river, I nearly blacked out. I shivered as I huddled deep in my sleeping bag, fully clothed, when camped at one of my study sites so I could take round-the-clock air and soil temperature measurements.

At times, the cold was sharp and brutal, as with the hair-washing incident, and at other times, it was dangerously insidious. Once, when lying on the tundra immersed in measuring microtopography (small-scale dips and rises) and the distribution of arctic white heather and other plants on hummocky terrain, I forgot how cold the soil was. I forgot that permafrost—ice crystals intermixed with soil—lay mere tens of centimetres below the ground surface. And I forgot to pay attention to the warning signs of hypothermia. It was late August, and by then, the High Arctic summer was gone. Early snowfalls had already dusted mountains across the fiord. Reddish-purple heather shoots and golden leaves of arctic willow, which grows prostrate along the ground, painted autumn colours over the lowland. Somewhere in the back of my mind, I registered that I was shivering, but—laser-focused on my work—I pressed on. I pressed on until my shivering became uncontrollable and my limbs were so stiff and chilled, I could hardly move them. Frightened into action, I forced myself to get up off the tundra and walk briskly—at first, stumbling over the hummocky ground—to base camp several kilometres away. The movement and my pumping blood warmed and loosened my stiff limbs, and my shivers became less violent and then receded altogether. I learned a lesson that day: I was not nearly as proficient a stress-tolerator as the plant I studied.

Here at Ravenhill, I continue to spade and trowel out weeds in my flowerbed near the base of the pine windbreak. The work goes slowly, my right palm sore from trowelling for hours in the same flowerbed yesterday. I stand up and, inexplicably, am filled with sadness, as though a precious relationship had ended, as though not a single person knew me, as though no one loved me—even though I know this isn't true; my family loves me. Then I realize why this feeling has swamped me.

Last evening after supper, Jānis and I mailed a letter and exchanged library books in Tatamagouche. We bought sunscreen and dental floss, as well as marshmallows and the oatmeal and chocolate chips we later used to bake cookies. It was Friday night, and the village's main street hummed with traffic. Fiddle music drifted from the Legion. I watched a young couple cross the street and climb into a gleaming red pickup with another couple already in it. The men sat by the doors, and the women in the middle, their laughter wafting out open windows. In that instant, I wanted—more than anything—to go on a date, to go dancing, to hear fiddle music drift from a community hall into a summer evening and feel the warm strength of a man's shoulder beneath my hand as we waltzed. I feel the same way now. Vilis and I have filled our lives so full of teaching, research, drawing, writing, chores, and driving our sons to and from games, lessons, and camps that we've left no time for dates, for dancing, for fiddle music floating into summer air, or for the warm strength of his shoulder beneath my hand as we waltz.

I leave the wheelbarrow half-filled with invaders and take my loneliness inside. I cook Red River Cereal for Jānis's and my breakfast, then gather shin guards, water bottle, and composure and drive my son to his soccer game.

June 25 (5°C, clear, calm, 7 a.m.)

This morning, our lawn is again a glistening sea of grass awash in heavy dew. I shiver. The air is fresh and cooler than I expected. As I step out briskly on the back road, a chipping sparrow hails the new day with its metallic trill. As I pass through the meadows south of the house, a snowshoe hare in summer pelage hops onto the road, then freezes when it sees me. The hare's underbelly and feet are wet and white, its ears, edged with black, and its body, tan streaked with darker brown. What does this hare see when it looks at me? Not colours, I know, but perhaps a strange, pale-skinned, looming shape? I creep forward. The hare shakes its ears and twitches its nose. I creep closer. The hare looks more nervous with each of my approaching steps until—its circle of safety breached—it bolts into roadside grass and alders.

In the backroad woods, dew drips from branches, and the new needles of balsam fir and hemlock are lime green lace edging the olive and forest green of older needles. Sunlight slants through the trees, highlighting tree trunks in stark detail and revealing mosses and lichens sprouting from bark. On the forest floor, fertile fronds of cinnamon ferns thrust upward like rusty spears from emerald fountains of vegetative fronds.

Even before I reach the top of the back road, where it ascends a rise, I feel the touch of sun-warmed air on my bare legs, although I'm still in shadow. The gentle heat is like the approach of something better, first only a hinting touch, then embracing me with its reality. What more will this day hold for me? Already I've been blessed with birdsong and a wary hare, with vivid new conifer needles and richly hued ferns. These blessings from nature take me away from swirling introspection, to which I am prone, and connect me with the rest of Creation. They link me with God.

In his book of Psalms, King David wrote, "The earth is the Lord's, and everything in it..."[22] If this is the case, and I believe it is, then I walk on holy ground, with the evidence of God's creativity all around me.

Once, decades ago when I was taking a university introductory Biology class, the instructor responded to a student who asked him which theory for the origin of life—creation or evolution—was correct. The instructor, a young professor who had completed his doctoral research in the field of botany, replied that he didn't believe the two theories were necessarily mutually exclusive.

The same question cropped up during my subsequent years of undergraduate studies, which included courses in Taxonomy, Anatomy, and Physiology that opened my mind to the wonders of the earth's creatures, and courses in Evolution and Ecology, subjects that place living organisms within an overall context of time and place. For three summers, I worked for a paleontologist and experienced firsthand the compelling evidence for the theory of evolution as I meticulously chipped away stone and watched fossilized fish skeletons appear before my eyes. And yet, throughout those years when I was immersed in science, my heart sought God until I was found by him. I respect science and all it reveals, but I concur with the instructor of that introductory Biology class. Why must creation and evolution be mutually exclusive? Madeleine L'Engle wrote, "Science, literature, art, theology: it is all the same ridiculous, glorious, mysterious language."[23] Perhaps evolution is simply God's method of creation.

At the top of the back road, I turn to walk home, my steps, light, my night's sleep too long and deep to allow me to walk the loop before Vilis leaves for work. Overhead, two male winter wrens duel over territories, their bubbling, sweet-toned voices clear and challenging in the cool air. For an instant, I'm one with them, singing high in the treetops, my music floating into a sun-streaked forest on a summer day.

June 27 (12°C, overcast, light S wind, 7 a.m.)

On April 20, 1840, Henry David Thoreau—famous for his musings on living the natural life at Walden Pond—penned this entry in *Journal*: "An early-morning walk is a blessing for the whole day."[24] This morning, as I set out on my early-morning walk, a deer bounds through a blackberry patch below the young pine windbreak, the underside of the doe's tail a flap of white that lifts like laundry in the wind. This outing, intended to awaken the senses, has already served its purpose. My ears attune to an eastern phoebe rasping out its husky song, a thick contrast to the monotone of junco and the quick clarity of Tennessee warbler. My eyes drink in bright wildflowers in the parkland meadows, where yarrow and chickweed bloom white, vetches bear purple flowers, and hawkweed raises yellow composite heads that will remain closed until the sun, not yet high, coaxes them into spreading their petals.

In the woods alongside the back road, a broken trembling aspen twig lies on muddy gravel, its grey-green leaves rounded and tipped in the same generous curves as those of a valentine heart. A striped maple sapling at the roadside bears broad, soft leaves—a hand's width of plant fibre that makes an excellent substitute for toilet paper in an emergency, as my family has found on more than one occasion. The time is too early for black flies, but mosquitoes bob around me, humming and landing on bare limbs, so I swat as I walk. Yesterday's wind, although strong enough to break aspen twigs and iris flowers, at least gave respite from blood-sipping insects. As the backroad forest encloses me in its soaring peace, I feel my spirit quiet. What would I do without this borrowed woodland? Without my walks and runs? Without Ravenhill?

After my family moved to Nova Scotia seven years ago, we lived for ten days at the Rainbow Motel in Truro, surrounded by pavement and the heat that bounced off it. Then we rented a modified trailer house in Lower Onslow, beside Highway 2. After three months of hearing transport trucks roar past in the night, we moved to a small house in nearby Belmont. Although another paved road passed in front of the house, the Chiganois River ran behind it, complete with a tangle of alders, raspberries, grasses, blueberries, young hardwoods, and the occasional spruce tree growing on the river's bank.

In the next nine months, during which Jānis was born, I walked a path through that riverside tangle many times. And on evenings and weekends, Vilis and I, with the boys in the back seat of our car, searched for a place of our own where we would be surrounded by nature. We drove every road within a thirty-five-minute radius of Truro, viewed properties with real estate agents, then gave up on real estate agents. Finally, in desperation, I wrote a newspaper ad:

> "CAN YOU Sell Us Our Dream? Wanted land
> with or without house, 20-100 acres or more,
> 5-20 acres cleared, rest forested. Private setting
> absolute must, within 35 minutes drive of Truro.
> Prefer long lane, shade trees, and view."

The ad ran for two weeks, during which Vilis and I received twenty-five telephone calls but found nothing we wanted, or in some cases, that we could afford. Two more fruitless weeks passed, causing us to sink deep into despair. One morning, hoping to encourage us both, I said, "Maybe someone will phone us out of the blue." That evening, a woman called. She'd been given our ad by a friend and was interested in selling a property south of Tatamagouche.

The next day, on August 12, 1991, we drove over Nuttby Mountain, past Earltown and The Falls to Matheson Brook Road. There was no sign, just a mailbox opposite a narrow, shaded lane that followed a brook. Roadside trees and shrubs leaned into one another overhead, forming an enticing tunnel. Vilis threw me a glance. "This looks interesting."

We drove over an old wooden bridge, up a steep hill, and into a grassy yard surrounding a squat brown house kept from loneliness by the presence of a tall shed, old apple trees, and windbreaks of young spruces and pines. The kitchen in the house was dark and gloomy, but the living room and upstairs were bright and warm. It was as though two sisters lived there—one dour, one cheerful—and fought each other to maintain their separate domains. We left the property by a longer road that passed through meadows south of the house and then through hemlock forest. When we paused briefly so I could nurse my baby, we heard coyotes howling. For nature lovers like Vilis and me, the howling seemed like a sign.

After that first viewing, we bought air photos of the six-acre property, which had been part of a seventy-eight-acre farm divided into two properties three decades ago. We came again in the heat of mid-August and explored parkland and meadows surrounding the house. Then Vilis came twice on his own and walked beneath tall hardwoods, fought his way through dense spruce saplings, and discovered an old wood road, a grove of sugar maples, and moss-covered rocks heaped along the edges of old fields. He arranged a building inspection, deed search, land survey, and lastly, negotiations with two sets of owners, since we'd decided we would only buy the house if we could get the whole farm. On September 10, 1991, less than a month after first seeing the property we later named Ravenhill, my family moved to a new life.

Some dreams, like Ravenhill, you can buy. Some, like the invitation to present an exhibit of pen-and-ink drawings, fall in your lap. And some, like the weed-infested flowerbed at the base of the pine windbreak, which I'm intent on reclaiming, you have to fight for and ache for because you know that without them a light in your soul will flicker and die.

As a writer with kids underfoot, I've learned much about fighting and aching for dreams. Before I had children, I daydreamed of delicious words that laughed and growled and sang in perfect formation. I brooded for hours, even days, over the perfect lead, the perfect mood, the perfect ending. Then I had a baby, and gone were the days of dreaming and dawdling. I changed diapers a dozen times a day, nursed my little one every two or three hours, cuddled him, cooed to him, and dealt with colic. The only time I could even think about writing was when Dainis napped, but even that was touch and go, each word scribbled furiously under the unbreakable deadline of my infant son's terrifying hunger. Still, I was a writer, and I had a *Ranger Rick* article about Dall sheep to finish, so I pushed myself to the limit, writing faster, focusing sharper, straining each nap's limits until the writing bubble burst from the impact of shrill baby cries. I completed the assignment, but my son wasn't the only one crying as I typed the last sentences. He was six days old. After that, I didn't look at my computer for eight months.

Fast forward several years to the day *Ranger Rick's* editor was scheduled to phone me at 9 a.m. to discuss two assignments, one article about arctic foxes, and the other, about hibernation. By 7:30, I had both my sons up, fed, and dressed. Just before 9, Jāni wanted his nap. Perfect! At three minutes to the hour, Daini asked for a snack, and I decided it was better to indulge him than to have him bother me when I was on the phone. He requested peanut butter toast and cocoa, so with no time to negotiate, I dumped a cup of milk into a pot, turned the burner on high, and then furiously mixed cocoa and sugar into the milk before popping a slice of bread in the toaster.

The phone rang, and I rushed out of the kitchen and around the corner into the study to answer it. The caller was Gerry, the editor. As we discussed possible animals to include in the hibernation article, a small voice from somewhere in the background quavered, "Mama, my toast is burning."

I turned away from the phone and called to the kitchen. "That's okay, Daini. I'll make you another piece when I'm finished my phone call."

"But, Mama, there's *fire*!"

I turned back to the phone. "Ah, Gerry, would you excuse me a moment?"

"Sure."

In the kitchen, flames shot out of the toaster, scorching the underside of the over-counter cupboard. I unplugged the toaster, forced up the lever that had failed to pop up, blew out the flames, and tossed the toast across the room into the sink. Intending to run cold water over the smoking, charred bread, I ran past the stove and heard a sizzling sound and smelled burnt chocolate. A small puddle of goo bubbled in the bottom of the pot I'd set on the burner. Trails of cocoa trickled down the pot's sides into the catch basin below the electric stove element. I tossed the pot in the sink, ran cold water over my son's "snack" and promised him another after the phone call. "Sorry for the interruption, Gerry," I said smoothly on my return to the study. "My son's toast burned."

His voice rich with laughter, Gerry replied, "Well, these things happen."

They do indeed, but my writing dream lives on because I fight for it and ache for it, and because I know that without writing, a part of me will be barren.

June 30 (15°C, clear, calm, 7:00 a.m.)

Calm morning air promises thick heat later today. In the meadow south of the house, grasses bear silver-bronze panicles of inconspicuous flowers. Goldenrod stands knee-high and lush, and blackberry canes wear frills of white blossoms. The richness of summer is upon us.

Warblers fly through the cool sea of the backroad forest. Air jumbled by my movements and laced with my body heat and exhaled carbon dioxide lures a bobbing armada of mosquitoes to their mobile feast—me! On the forest floor, an ovenbird rustles dry leaves before flying upward to perch on a hemlock branch. Is this the mate of the male singing nearby? Our Nova Scotia dawn chorus is still strong, still filled with melodies that together comprise a vibrant polyphony. The musical performance has daily variations dependent on which singers are in voice, but admission is always free. I simply step out my door and listen.

On my return to the yard, I pause beneath an ancient, twisted apple tree, my eyes unconsciously seeking a rectangular outline that is no longer visible on the lawn. We buried Skyree here three springs ago. A Siberian husky/Samoyed cross, my Sky was big and beautiful, with red-gold fur. He pranced when he walked, and when young, ran like the wind.

Sky was a puppy when I took him with me to Heart Lake Biological Station in the Northwest Territories, where Vilis and I fell in love. A snow dog, he cooled off on ice in deep crevices of an escarpment that looked out over tens of kilometres of boreal forest between the research station and Great Slave Lake. A year later, I took Sky with me to Alexandra Fiord, where he sprinted over the High Arctic tundra, playing touch me if you can, and alerted me to a polar bear treading on the fiord ice. Sky was eight when Vilis and I moved to a cabin beside Kluane Lake in the Yukon. My surrogate child for so long, he learned to share my affection with the newest member of our family, baby Dainis. Sky was twelve, his muzzle frosted with white, his hips sore from arthritis, when we moved to Ravenhill, yet he still pranced when he walked and occasionally romped with us in the yard, displaying some of his old speed.

The decision to euthanize my beloved canine companion—almost fifteen years old and filled with cancer—

was the most painful decision I've ever made. Vilis spared me the agony of taking Sky to the vet and brought him home again, wrapped in his old blue flannel blanket and positioned so that I wouldn't see his right eye bulging from the tumour behind it. Sky looked like he was sleeping. I touched his soft fur gently before we placed him in the grave that Vilis and the boys had dug earlier under the apple tree's boughs, where so often in the heat of summer Sky had lain in the shade, his body stretched out luxuriously, his red-gold fur lifting and tossing in the wind. I cried when we buried him, and so did Vilis, while our sons watched us with still, searching faces.

Now it's evening, hours after my morning walk and bittersweet remembrances. Today, time stretches behind me as a kaleidoscope of walking and weeding, of kneading bread and washing floors, of hanging laundry and planning a drawing, of checking the garden with Jānis and forming bread loaves while the Union Jack came down for the last time in Hong Kong, of washing dishes and biting into crusty steaming bread, of drawing a piece of art and wading in Matheson Brook, of cooking broccoli soup and coaching Dainis's baseball team, and of eating one last slice of fresh bread before bed. This kaleidoscope is one day of my life, one snapshot of a simple existence wherein each shifting, overlapping piece is vibrant with its own beauty.

July 2 (14°C, clear, calm, 5:45 a.m.)

Yesterday, torrential rain chased me from weeding the front flowerbed. Heavy drops pounded the porch deck, bounced up from our car, and streaked the air with silver. Fifteen minutes after the downpour began, the deck was quiet. Our gardens received only a sip when what they need is a long drink. The frequent rains of May are long gone, replaced by hot cloudless days that suck moisture from the earth.

This morning, the rising sun is an orange orb above curtains of white mist rising from Matheson Brook. Cool dawn air, heavy with moisture despite the lack of precipitation, is as fresh as a drink of mountain water. An American goldfinch wings above the lawn, its undulating flight suggestive of a lemon-tinted blossom buffeted by the wind. Closer to hand, a male ruby-throated hummingbird hovers in front of me, its wings a noisy blur, its iridescent green back and red throat flashing brilliant colour when caught in sunlight. I'm reminded of a day when I stepped onto the porch while wearing a purple sweatshirt with pink flowers on it. A nectar-hungry hummer zoomed in and hovered in front of my chest, seeking the flowers' corollas into which to insert its bill. Alas, there were no real flowers, only fabric deceit. The avian emerald hovering inches from my body mesmerized me, as does the hummer hovering in front of me now.

Unlike birds that hover briefly while hunting for food, hummingbirds are true hoverers. If the air is calm, they can remain suspended in the same spot for as long as they choose. A whopping thirty percent of a hummer's body weight is devoted to breast muscles that power this tiny bird's wingbeats during its high-speed, maneuverable flight and hovering. In addition, a hummingbird has flexible shoulder joints that allow its wings to move in a sideways-flattened figure-eight pattern that creates lift on both forward and backward strokes while hovering. This pattern maximizes energy efficiency by sweeping the wing through the path of air molecules already pushed into motion by earlier wing strokes.[25] A ruby-throated hummingbird performs seventy-five or more wing beats per second while flying forward, sixty-one per second during reverse flight, and fifty-five per second while hovering.[26]

Each spring, I eagerly await the first song of robin, song sparrow, yellow-rumped warbler—of each migratory songbird species that returns to Ravenhill. Yet I always listen expectantly for one certain sound, not a song, but a sound that, nonetheless, announces the return of a vivid jewel in the crown of eastern North American avifauna. At the first thrum of super-efficient wing beats, I duck into the house and announce to Vilis and the boys, "Hummers are back!"

July 23 (5°C, clear, 7 a.m.)

Home again after two weeks of visiting Vilis's family in Ontario, I cling to a veery's descending flute-tone song and mourn the time away from Ravenhill. Late summer's yellowed green pervades the landscape, and our lawn is brown and crisp. In the meadows, seed heads weigh down grasses, but in my garden, beans and peppers have been decapitated, likely by deer. Seen afresh, our house looks old and worn. Still, not all is to be mourned, for three newly hatched chicks chirp loudly in the coop, and pansies in a flower box bloom profusely beside the feed room door, their open faces welcoming us home.

Four decades and a year of life have taught me that it's often small things, like newly hatched chicks or a box of pretty flowers, that help me get through hard times—times of change and adjustment, or even times like today when I return to a beloved place and find that it has altered in my absence. Life is rife with change, whether initiated by us or occurring in the world around us. All we can do is meet new circumstances head on, although we may wish to tuck something reassuring in our pocket.

Here's a story about change, and about unexpected love that blossomed within that change. After I finished my arctic white heather studies, I put the possibility of more research on hold so Vilis could pursue his goal of a career in zoology. It was during this hiatus, and while I assisted Vilis with his skunk studies, that I had an epiphany. I realized that I don't live for science. This new understanding came about as a result of going for walks with Vilis, during which he mentally solved mathematical ecology equations while I searched for words to paint images of the trees and fields around me. On those walks, I wasn't thinking science; I was thinking writing. As intriguing as High Arctic plants and striped skunks and birds might be, it was clear to me that I didn't possess the life-gripping passion for science that I believe is a prerequisite for being an effective research scientist. So, where did that leave me? If my life wasn't to hold a career in science, what *would* it hold? I began to dabble in writing.

That's the change part of the story, yet life rarely proceeds from A to B without a few curves in the road. The unexpected-love part of the story began with a dream, Vilis's dream, or perhaps more accurately, with a continuation of his dream. A devotee of wilderness and the North, he yearned for a chance to live and do wildlife research in the Yukon. And as sometimes happens when one doesn't give up on dreams, his came true, in the spring of 1987. Snowshoe hare biologist wanted at Kluane Lake: was he interested? Definitely. But what about me? Did I want to go? The lure of the North was nearly irresistible; yet, being a solitary type, I had qualms about living year-round at a research base. Still, a rustic cabin on the shore of a glacier-fed lake did sound like the ideal haven for my planned metamorphosis from biologist to writer.

Decision made, we packed our bags and directed mail to Mile 1054 Alaska Highway. Once at the Arctic Institute of North America's Kluane Lake Research Station, Vilis and I set about personalizing a brand new fifteen-foot square log cabin. We added a deck, built a loft, painted a clear finish on the plywood floor, and covered second-hand furniture with bright afghans and patchwork throws.

Yet something was missing. The cabin was so new, so bare, so overpowered by turquoise Kluane Lake and the jagged St. Elias Mountains. And the base, with researchers coming and going in flurries of biological and geophysical activity, had an air of impermanence about it, like that of a summer camp. Despite my love for outdoor adventure, I needed something to balance the rugged landscape and artificial community, something small to ground me during that time of change in my life. White lace curtains, perhaps? A flower box full of bright blossoms? Now that was an idea. I closed my eyes, and a vision of riotous beauty leapt to life. In the vision, petunias' velvet throats and rumpled skirts blazed against our cabin's pale gold logs.

I ordered petunias from Whitehorse, but pansies came instead. *Oh, well,* I thought, *pansies are...nice.* While researchers hammered and drilled and painted research materials—some designed to protect supplies from marauding grizzlies—I built a flower box. The next day I screwed it onto our cabin wall and planted the pansies, so leggy and sprawling that I despaired they would ever amount to much. Within a few days, however, wedding-cake buds unfurled into blossoms—first white, then navy, then burgundy and gold, then deep red, pale yellow, rose, purple, lemon yellow, coral, and finally, orange. I gaped, chagrined that I had ever despaired.

Each day I watered the flower box's occupants and removed wilted blossoms, savouring the exquisite feel of velvet petals before tossing them in a tumble of bright colours into the willows by our cabin. While at my desk, I peeked out the cabin window to be cheered by the pansies' vibrant hues. I often read on the deck, just so I could glance up now and again to bask in the unexpected success of my idea. And each time I left the cabin or returned, the faintly nodding blossoms bid me farewell or welcomed me once more.

I came to depend on those pansies. On days when I couldn't put two coherent sentences together, they encouraged me. At times when my solitary self was swamped by forty researchers, assistants, and base personnel, the pansies were my refuge. And when the occasional patronizing academic made me wince ("So, you're just the wife?" "Ah, yes, your *stories...*"), I retreated to white lace and pansies. They were kindness and compassion; they were home in a place that wasn't my own.

Late August brought dancing lights in the night sky and the risk of a killing frost. I couldn't bear the thought of shriveled, blackened pansies, not yet. Each evening I draped a piece of flannel sheeting over my beloved flowers, and each morning removed their protective cloak. Then, one night, I forgot. In the morning, my pansies sprawled lifelessly over the flower box, their leaves stiff, as though carved from wax, their stems the same translucent green as frozen English cucumber. "Pull out pansies," I wrote on my memo pad, not wanting to see them like that. Just before noon I stepped outdoors to do the task—and stopped. Stems arched, and flowers flared to greet me. Not only had my pansies survived their plunge below zero, they had rebounded into all their endearing loveliness.

Snow premiered on the last day of August. My pansies wore its cold white lace with grace. September blustered and blew and shone its way into October. The last of the summer research crews departed for southern cities while three more storms dropped transient snow and occasional lows of –10°C. Each morning I looked at my pansies—wilted, heads hanging—and thought, *This is it. They're dead.* But they never were. Each afternoon they lifted velvety faces to sun or cloud, sometimes rain or snow. Long after native wildflowers slept, my pansies still warmed our cabin wall. Of course, I knew they couldn't last forever. Still, when the deep-freeze descended in late October, I was saddened. My valiant pansies could lift their faces no more. They never did become black and shriveled, they just didn't wake up again.

Ten years later, and now living 5,000 kilometres away from the Yukon, I still need pansies. Each spring I smile as I plant them, and I touch old wood lovingly, for I brought the flower box with me. Its sides are grey now, its screws, rusty, and cracks split open its seams. Still, it was this same flower box—when its wood was new and golden and it housed unexpectedly charming and valiant pansies—that became my anchor in a world of jagged peaks, turquoise lake, research base crawling with summer personnel, and uncertainty and self-consciousness as I traded my life as a biologist for that of a writer. I came to love those pansies in a way I never knew I could.

Who can say where love will lie, or how it will come to be? Love is as ungraspable as the universe and as profound as childbirth. It beats its own path to the core of our being, and once there, touches and expands that core, leaving us more than we were and guiding us toward all we can be. To love is to glimpse the face of God, even when the recipient of that love is as seemingly insignificant as a box of pansies.

July 26 (8°C, mostly clear, calm, 6:20 a.m.)

This morning, mist rises from the brook and meadows and wafts around rhubarb leaves. Dew decorates the lawn with lacy spider webs. Before I rose from my bed, I heard the jumbled swell of a dozen bird songs. Now the voices of black-capped chickadees and white-throated sparrows lift clear and strong around me as I again fight opportunists. This time I'm working in a strip of soil west of the house that would be, were it not for the pressures of teaching and drawing, the strawberry bed of my dreams. Instead, it's a bed of weeds.

The soil is dust in my hands. According to the postmistress in Tatamagouche, the only real rain our area received during the past month was a thunderstorm last weekend. I see this dearth of moisture in my stunted cucumbers and in the two withered berries on each of my high-bush blueberry plants. The latter are almost smothered by weeds at the far end of this neglected garden. Yesterday I discovered they were still there, still alive and reaching for the sun. So were a grape vine, kiwi vines, and domestic blackberry canes that I thought the past winter, with its frequent thaws and lack of snow, had killed. I freed and watered the grape and blackberries, and today will tend the blueberries and train the kiwi vines to poles Vilis erected months ago. In caring for them, I'll free myself from the guilt of neglect.

By 10 a.m., the sun's heat and my back have defeated me, the former scorching, the latter wrenched and throbbing. I leave the weeds, postponing freedom, and harvest tart Montmorency cherries with Jānis, whose eyes sparkle in anticipation of pie.

Despite weeds, deer depredation, and mid-summer drought, this small patch of earth we call Ravenhill provides great bounty. Plant a seed or fruit tree, care for it, and in the fullness of time pluck its tasty delights. Whenever I visit friends who share my passion for growing things, we end up in the garden, discussing what's growing well and what's not.

The word "garden" is derived from Old North French *gardin* rooted in Proto-Germanic *gardan*, meaning something along the lines of "an enclosed yard." Surely one of the most beautiful words in the English language, "garden" equates with good food, caring for soil, and hours spent outdoors. It equates with hard work and getting hands dirty. It equates with *joy*.

July 27 (17°C, mostly clear, calm, 8:30 a.m.)

The three new chicks are dead, killed in the night by an attacker that ate one's wing, the head of another, and devoured completely or dragged off the third. No hens or roosters were harmed. What was the killer this time? Skunk? Weasel? Another raccoon? At 4:30 a.m., Vilis found where the predator had dug its way into the coop, which has a composting floor; however, he didn't see the culprit. I'm thankful we didn't buy Meat Kings this summer, since they also would have been in the coop—thirty-five defenceless chicks three weeks old and no longer cute in bare pink skin and rough new feather quills.

Despite the fowl murder, this morning feels like a gentle hug. Vibrant bird voices—red-eyed vireo, black-throated green warbler, dark-eyed junco—seem filled with expectation or perhaps the intense satisfaction of knowing that the sun has once again risen to bathe them in its brilliant warmth. I once read in a book (I can't recall title or author), that to foster a positive outlook on life we should surround ourselves with people who make us feel good. To this I add that we should surround ourselves with landscapes and living creatures that make us feel good. Human friends may be fickle, move away, or get misplaced in the busyness of living, yet every day rivers run and hills swell against the sky. Every day, bird voices ring out and there's land to explore, whether city park or mountain forest. The earth and its creatures await us. We have only to open our eyes to see, and our ears to hear.

Once I went for a walk on a mild overcast winter afternoon when I couldn't write anymore. My mind was filled with a leaping, tumultuous crowd of sentence fragments that heckled and tormented me, refusing to drop neatly into useful prose. Oblivious to the world around me, I walked two-and-a-half kilometres before dragging my mind off the scrambled-thought carousel and setting my focus outside myself, onto my surroundings. I became aware of freezing drizzle carried on a north breeze that peppered my right cheek. Dead ahead, a farm field rolled gently downhill to a farmyard, its snow-covered knolls like immobile white waves on a frozen sea. To the left, within this field and backed by a thick line of grey shrubs growing along a drainage course, a small pristine hillock mounded in a gentle curve. That snowy mound struck a chord in me, and as I gazed at it, my rioting thoughts settled into peace.

Without that perfect white curve of snow, I wouldn't have gone on. I wouldn't have walked farther on a quiet country road on a moody afternoon. I wouldn't have seen a red fox that trotted across a winter field, or found its fresh tracks, some lightly stained with soil, crossing the snow-patched road and then meandering up a white hillside. I wouldn't have lifted my eyes to the hills, their peaks shrouded with pale grey cloud that sagged into forest and onto hill farms, brushing them with mist. I wouldn't have noticed dark alder catkins hung like lace against the backdrop of those clouds. Or burdock seed heads like tiny pompoms held aloft on winter-brown stalks in a weedy ditch. The snow-covered knoll acted as a trigger that cast me into the realm of awareness and opened my eyes to the riches around me.

It's too easy to forget that I don't walk this earth alone, that I'm a member of a much larger and complex body of life that I desperately need not only to survive, but to thrive. One captivating sight led me to others. Tumultuous thoughts gave way to peace that led to quiet joy. Annie Dillard wrote, "I walk out; I see something, some event that would otherwise have been utterly missed and lost; or something sees me, some enormous power brushes me with its clean wing, and I resound like a beaten bell."[27] A snowy knoll swept away my thought chaos, freeing the bell within me to ring out its bright reverberations over the northern Nova Scotia countryside.

Who can say where we find ourselves? In the air? In the soil? In the lace of alder catkins hung against a backdrop of cloud-shrouded hills? In the smooth trot of a red fox across a snowy field? That winter day, I found myself in a small white knoll in a farmer's field. On the return portion of my walk, I photographed the knoll so I can look at it at my pleasure. When I'm lost in unruly crowds of badgering thoughts, I seek again its perfect gentle curve.

August 11 (15°C, clear, SW breeze, 7:10 a.m.)

A kingfisher chatters its rattling call overhead. Two deer, a doe and fawn, stand alert—heads lifted, ears cupped—beneath an apple tree in the meadow beside the back road. The deer stare at me even more curiously than I stare at them, but they don't bolt. Goldenrod blooms all around me, its blossoms the colour of lemons held on green stalks amid the gold of ripened grasses. Cool air touches my face as I approach the woods along the back road. At the forest edge, aspens shake out their trembling whispers, their trunks clean and white, their leaves, green coins high above me.

How I missed these walks! For the past two weeks, my sore back—injured while gardening—complained bitterly at the jolt of placing one foot ahead of the other. Yesterday, however, I crossed a brook, scrambled up and down hills, strolled a windy beach, and for the first time in a decade picked up my racquet and played badminton. The court was our sunburned lawn, the gymnasium roof, the sky, and the opposition, my nine-year-old son. This morning, I walk with confidence, healed.

When I reach Balmoral Road, sunlight flares in my face and bounces off the pavement. Having risen too late to walk the loop, I retrace my steps, squashing deer flies in my hair—two, three, five, seven, more. The kingfisher chatters again, and the aspens greet me once more. At the woods' edge, backlit bracken fern glows with green fire, as if the season were still spring, as if autumn would never come, as if the lush fern fronds would never turn brown and wither.

I water our chickens, then help Dainis and Jānis transfer frozen insects from plastic bags into vials. These are quarry my sons stalked and netted during the past two weeks. "Mom!" they yelled, running to me. "Look what I caught!" Grasshoppers. Bees. Shiny beetles. Leaf hoppers. Wasps. Butterflies. Their prizes will expand the entomology teaching collection at the Nova Scotia Agricultural College.

In mid-morning, sunshine warms my back, and pea pods lie plump in my hands. The lushness of a summer morning surrounds me, but my heart holds a festering that remains unhealed because I allowed a friendship to be overgrown by seeds of tactlessness and neglect. Why do I sometimes say the things I do? Why do I sometimes do the things I do? Why do I find it so hard to explain, to admit I was wrong, to just come out and say it? Two pine siskins, wee and striped, land on pea stakes in front of me and then fly. I wish my hard, painful feelings would fly with them and leave in their place acceptance, understanding, and wisdom. There are times in life when I don't like myself very much. My drawings, now hanging in the 1889 Gallery, have received accolades, but I would trade all of them for one damaged friendship restored to full and brilliant blossom.

After harvesting the peas, I snap off sweet basil and broccoli, then plant bean seeds where the deer—perhaps the doe and fawn I watched this morning—devoured my green peppers. Sowing beans at this time of year is chancy. More than anything, the seeds are an offering, a repentance, a way of fulfilling my need to make a clean start somewhere.

Hours later, in late afternoon, Dainis, Jānis, and I stroll down the steep front-road hill, its gravel overgrown with grasses and goldenrod. Heat pervades the air, the temperature, 27°C. We're on our way to the brook for a cooling dip but take time to pluck red treasures from thorny canes and pop them in our mouths. The flavour of warm raspberries bursts onto our startled taste buds.

From the bridge, we see brook trout swimming cautiously in the pool under the deck, their freedom restricted by low water. Most of the trout are fingerlings, but at least a half-dozen are twenty centimetres long. Upstream from the bridge, a black-and-white dragonfly is a miniature bush plane skimming over a liquid runway. When we climb down the bank and into the brook, the fish scatter, and so do water striders. As the long-legged aquatic insects sprint for the pool edges, their feet dent the water surface, creating white-hot flashes of light on the pool's blue, reflected-sky surface.

The water is cooler than I expected. Dainis and I giggle as we summon courage to plunge into its depths. Finally, I count one—two—three, and he and I hold our noses and dunk. Jānis is already wet, as Vilis would be. After more refreshing dunks, I climb back up onto the bridge to bask in the sunshine, leaving the boys to play in the pool. Dainis soon joins me, but Jānis moves downstream.

The water striders that scattered when my sons and I entered the pool now venture back to midstream and tap out ripples with their feet. I hang my head over the edge of the bridge and watch them. One strider chases another strider from its space but allows a third strider to remain nearby. Is the banished strider a competing male? Is the accepted strider a female?

In the early 1970s, animal behaviourist R. Stimson Wilcox first interpreted the ripple dances of water striders. His studies revealed that male striders tap their forelegs, midlegs, and hindlegs on the water surface to create wave patterns. Those wave patterns attract female water striders, induce the females to copulate, signal copulation, and signal copulation's end. The male striders also use their midlegs to produce aggressive signals that Wilcox thought indicated defence of the striders' territories.[28] In later studies, Wilcox timed the striders' signals and reported that male striders dance ripple messages at two different frequencies: three to ten waves per second, and eighty to ninety waves per second. He determined that the fast strider dance is that of an aggressor warning other males away from its space, whereas the slow dance is that of a "lover" enticing females to mate. Female water striders only dance the slow dance.[29]

I look for fast dances and slow dances but instead am captivated by interference patterns created when the striders' ripples collide. Undisturbed ripples play out as rings of brilliant light on the brook's shadowed floor. When ripples collide, the light rings smash into each other and explode into shimmering, fading segments. And then there are the shadows of the striders themselves. These are collections of thin legs and sextets of round blobs that are shadows of the *dents* that the striders' feet make in the surface of the brook. This past spring, Vilis and the boys investigated the physical property of surface tension in liquids. They laid paper clips and small sheets of aluminium foil flat on water. Both lightweight objects provided respectable illustrations of the cohesiveness of water's surface molecules, but neither could rival these striders. They walk on water.

My sons and I leave the brook, retrieve our mail, and climb the hill to the yard, where lawn grass crackles beneath our sandals. During the past two months, the hottest of the year, we've received only three brief showers. We have, however, had success trapping raccoons. The day after the chicks were killed, Vilis trapped a coon in the coop and drove it kilometres away before releasing it. We've had no predator problems since.

August 19 (12°C, cloudy, calm, 7:40 a.m.)

Dark blue clouds billow overhead. At last, we've had rain—no long soakers, but several quenching showers during the past few days, including one this morning. The air is noticeably cooler. Yesterday, for the first time since winter's end, my bare hands felt chilled.

Robins and goldfinches flush from the meadows beside the back road. I hear a blue jay's strident call and a white-throated sparrow's whistled song, the latter sounding "uncertain." Perhaps, the sparrow is a youngster trying out its voice. White-throats are a handsome species, with their dramatic black-and-white-striped heads, splashes of yellow in their eyelines, and white throats surrounded by solid grey. Intriguingly, these small songbirds show two different colour morphs. One morph has a white eyeline behind the yellow splash, and the other has a tan eyeline.

In 1975, Vilis spent a summer working for an ornithologist—Bruce Falls—who studied white-throated sparrows in Algonquin Park in northern Ontario. The aim of the research was to test the importance of birdsong as a means of keeping intruders out of breeding territories. Each morning, my one-day-to-be-husband rose at 4:30 a.m., slathered on DEET to deter northern Ontario's voracious black flies and mosquitoes, and headed into the boreal forest. Several times during the day, he recorded songs of territorial male white-throated sparrows to determine how often they sang. Another assistant who was a gifted guitarist came up with the idea to use a pitch pipe to determine the exact notes sung by the male sparrows. It turned out that each sparrow had a unique song that consisted of three notes. In some of the songs, the three notes ascended in pitch; in others, they descended. Regardless of whether the notes went up or down, the pitches and pattern were consistent for individual sparrows. In essence, each white-throated sparrow had a vocal fingerprint.

The next research step was to map territories of singing male sparrows. A territory was the area that a male defended against other males and within which it mated and raised young. To map a territory, Vilis found one of the male sparrows whose song he had earlier recorded, then played a strange male's song. If the resident male swooped at the playback recorder—the equivalent of attacking an intruding male—Vilis knew he was within the resident male's territory. "It was kind of fun," he said of the sparrows' swooping attacks. If the resident male flew only so far and then stopped, Vilis knew he was outside its territory. Thus, he was able to map territory limits.

After territories were mapped by Vilis and other assistants, the research team live-trapped male white-throated sparrows and removed them from their territories, holding them temporarily in cages. Then Vilis (or another assistant) performed three tests within territories where the resident males had been removed: in some of those territories, Vilis played the absent male's song, in some, he played a strange male's song, and in some, he played no song. When no song was played in the absent male's territory, it took only half a day for another male to move in; however, when the resident male's song was played, several days passed before an intruder invaded the territory. After an intruder did move in, the resident male was returned to his hard-earned territory and made quick work of evicting the intruder. So, a white-throated sparrow's song is key to defending its territory, and the same has been shown for other songbirds.

Now I wonder, what is my territorial song? What do I sing from the heights to proclaim the start and finish of the living space in which I raise my brood of two? How do I defend Ravenhill? Do I need to defend Ravenhill? Do I need to speak up for it, to extoll its beauties, to announce my keeping of it and the legacy it brings my family? The answer is yes, and the answer is words. The handsome, clean-cut male white-throated sparrow raises his bicoloured beak to the sky and whistles "oh, sweet Canada, Canada." I write, painting images of the world around me. Among white-throated sparrows, females of the "white-striped" colour morph also sing. Perhaps I'm one of those.

In the hemlock forest alongside the back road, raspberries growing in a clearing bear yellowed leaves. A roadside maple's foliage is tinged with burgundy. The season shifts toward autumn, toward shorter days and ever-increasing darkness that's like a veil shrouding the earth, a veil that moves in two directions, sliding progressively higher to cover more of morning and progressively lower to hide more of evening, until winter solstice, that time of the longest night when the sun creeps so late from its bed and retires so early. But the solstice is still months in the future, and colour-rich autumn will brace me for its darkness.

Balmoral Road is quiet. I hear only a gentle rustle of leaves and the drip of raindrops falling from trees. No cricket melody. No "clack-clack" of grasshoppers in mating flight. Those songs await the sun. A trio of northern flickers flushes from roadside shrubs. Where are they going? Where will *I* go today? What new thing will my mind discover? Each morning, sunlit or rain-soaked, is an open door, an invitation to explore and discover—if I see the door and choose to walk through it.

Albert Einstein was once asked to give a word of advice to a bright young physics student who suffered from an overwhelming sense of hopelessness. Einstein told the young man, "Be a loner. That gives you time to wonder, to search for truth. Have holy curiosity. Make your life worth living."[30] I think again of Mary Oliver's poem "The Summer Day" and believe that Oliver took the time to wonder, that she searched for truth, and that she had "holy curiosity." My hope is that I will do the same, steadily growing my curiosity until it glows like a blazing flame within me.

Balmoral Road's pavement leads me to gravel Matheson Brook Road, the third leg of my loop walk. Beside the road, fireweed blooms magenta on tall stalks. Shiny alder flea beetles crawl on alder leaves that have been chewed into elegant skeletons of netted brown lace by the beetles' larvae.

I make the sharp turn onto our front road, that corridor alongside Matheson Brook that's like a tunnel through overarching vegetation. Past the bridge, I'm in the open, with moisture from rain-laden grass filling my sneakers as I climb the hill. In the yard, my garden, as though in sudden awareness that only a month might remain before the first killing frost, pushes out lush vines of cucumber, squash, and pumpkin even as pea plants fade. A watermelon, plump and rounded, and several cantaloupes lie half-hidden among concealing vines and leaves. Jānis, who prizes the garden, checks on these treasures each day.

August 22 (11°C, overcast, strong SE wind, 8:05 a.m.)

Today, I'm restless, restless like the wind that blows scattered raindrops onto my lips and bare arms as I walk the front road on my way to collect mail. There is no joy in this restlessness, only the need to do and the feeling that I haven't done enough. Yet my days are filled with tasks—baking bread, washing laundry, caring for my gardens, raising my sons—but none of these tasks bring in cold, hard cash.

And thus, the quandary begins. What to do with my drawings now hanging in the 1889 Gallery? I've met with a man who would make, at his own expense, photographic reproductions of at least three of the originals and market hundreds of them as limited and unlimited edition prints across the country. He believes in bringing art to the masses. I've met with an artist who believes that photographic reproductions are not art at all and who advises me to do silk screens of my works, simple when dealing with black and white, and each one hand-pulled to make it essentially an original. I've spoken with friends who aren't interested in reproductions of any kind, only the single inspired original. The artist's coat I planned to hang in my closet after my show's opening is proving clingier than I expected. At times it tightens around me until the freedom with which I drew those pictures is squashed into panicked confusion.

After collecting mail, I stoop to pick up an empty cigarette package—one of many I've gathered this summer—and tuck it in my pocket. Too many people who drive Matheson Brook Road are rotting their lungs, as my father did. Fifty years of smoking roll-your-owns did his lungs in, then the rest of his body. He became a skeleton of himself, shaking as he walked in slow, uneven steps to the kitchen table with my brother Tod's arm around him. Dad was home from the hospital for a few hours and nauseous from chemotherapy, but he wanted to eat red Jell-O with fruit in it, one of my mother's frequent Sunday dinner side dishes. His hands bounced up and down as the spoon moved toward his mouth, but the glassy Jell-O slipped away from his trembling lips and fell back into the dish. "Ah!" he muttered in frustration, his voice old, raw, hurting. This man, who had farmed until he was seventy, only then handing the reins over to a son, could not even feed himself a spoonful of Jell-O. We—his family—sat aching at the table with him, our throats choked with the bitter helplessness of watching him die in slow steps before our eyes.

How is it that grief strikes so suddenly, a bolt of pain? I walk with a cat, and it slams into me. I pick up a cigarette package, and it stabs me. Where do I turn to escape or calm the raging lion that leaps from the darkness?

On the front road, lacy skeletons of mined alder leaves jolt me from my aching memories. It seems impossible that destruction can yield such delicate beauty, yet the evidence lies at my feet. I gaze at the leaves, almost dazed, caught in this image of loveliness arising from devastation like a phoenix rising from the ashes. What beauty rose from my father's death? What captivating elegance? Where is the bird of resurrection rising from his long-buried casket? I see none, and yet the brown lace of skeletonized alder leaves comforts me.

Near the bridge, chokecherries hanging from drooping stems are the dark red of venous blood. Pin cherries gleam like crimson beads shot through with light. As I climb the hill, sparrows and crickets cantillate. The earth tosses gift after gift at my feet, soothing my inner turmoil. Biophilia—the bond between humans and nature. Right now, it's biophilia that sustains me.

August 23 (16°C, rain, calm, 9:30 a.m.)

Today my family is looking for logs. Specifically, we're searching for two beaver-cut aspen trunks that slid off a trailer in the dark two days ago.

Last year, beavers felled giant aspens at the edge of the Amphitheatre. They gnawed off bark and branches and left thick, heavy trunks on the shore of the pond they created. We came along and saw lumber in those trunks: two-by-fours to frame new rooms in our basement, two-by-tens to span the breadth of the basement ceiling and brace the living-room floor, and wall boards to cover pink fibreglass insulation in the addition we built onto the kitchen three years ago after our cement basement was installed. The aspens were free lumber if we retrieved and milled them. For the retrieval, Vilis spent two afternoons and evenings with a tractor, chainsaw, chains, a borrowed log trailer, and occasional help from me. The milling is yet to be done.

We find the missing logs at the base of a spreading pine near the thicket of blackberries that arc like thorny ribs protruding from the meadow behind the garden. The logs are thirty and forty centimetres across. Other logs already home are forty-five centimetres across. It was only when I saw those aspen trunks lying like enormous pick-up sticks on the ground by the beaver pond that I truly appreciated the logging skill of beavers.

Loath to return to the house so soon, we follow the tractor's trail through the parkland north of the house and then down the old wood road that is a wood road once more. We kick soil and moss into tractor ruts, and Vilis shows us where the tractor got hung up on a slim pincherry tree, and where he scrambled in the dark to find the chain that jumped off the chain saw when he started cutting the pincherry out of the way. But he liked it out here, working with his hands, with wood, tractor, and saw instead of computer and keyboard. He tells us, "Getting those logs home was a lot more satisfying than solving an equation."

Light rain falls on and around us, easing the drought and filling us with peace. The boys and Vilis have umbrellas: black, navy-and-white, pale blue with white polka dots. I'm getting wet, but I don't care. The rain is warm.

When we reach the brook, past the Amphitheatre, it's obvious the beavers are gone. Their pond is shrunken, and the dam now looms over it, with jewel weed and bittersweet nightshade thick atop its ridge. We find a rotting beaver carcass in shrubs at the edge of the clearing where Vilis first piled the hefty aspen logs last fall.

This is what we wanted—the beavers gone—to save our sugar maples, yet the dead beaver, abandoned lodge, and bare mud and slimy logs and branches exposed by low water bring no pleasure or satisfaction, only sadness. Perhaps the drought did the beavers in, the water in the brook so low that even they couldn't maintain a pool deep enough for their needs. We hope our maples will recover and send forth more leaves next spring now that their roots are no longer encased in water-logged soil. For this hope I'm thankful, but still there's sadness, and a rotting carcass in the shrubs beside the clearing, its webbed hind feet curled like withered claws, its orange lower incisors curving like semi-circular chisels, its brown fur falling to the soil in disintegrating clumps.

We retrace our steps, and the rain is still warm on us, still as gentle as forgiveness for any deed done or undone. Dainis's rain pants, borrowed from Vilis, unroll at the bottoms yet again, so we stop to hike them up. As we pass our campsite, sudden yearning flashes through me, a yearning to be lying in a tent, immersed in a book, with light rain tapping its soothing song on the fabric above me—like when I was twenty and off on a fossil-collecting expedition in the Alberta foothills and it rained day after day. I was camped in a tent like that on the day my father died. I'd phoned my sister Alvina to check on Dad's condition, and then phoned again a few hours later. My father died between phone calls. That night I lay awake in my tent for hours, staring at the rust-brown walls and yellow zipper. The next day, I caught a bus for the long ride home. *I'm coming, Dad.*

Jānis abandons the shelter of his umbrella and is soon as wet as I am. In the parkland I find a fat green gall on a goldenrod stem, and we examine it with interest. There's no hole, so we know that somewhere within this inflated balloon of plant tissue a goldenrod gall fly larva is feasting and growing, safe within its stolen womb. Female goldenrod gall flies (*Eurosta solidagnis*) lay eggs in newly sprouted goldenrod stems, of which our land has an abundance. The eggs hatch, releasing larvae that eat plant tissue inside the stems. As the larvae chew, they release saliva that might mimic plant growth hormones, and in response, infested goldenrod stems overproduce new tissue, resulting in the formation of a bulbous gall, like the one I found. Larvae feed within the galls (one per gall) all summer and spend the winter inside them as freeze-tolerant grubs. In spring, the larvae construct pupae in which they metamorphose into adults that exit the galls and, after mating, initiate the goldenrod gall fly life cycle again.[31]

We leave the gall and pass our lost-and-found aspen logs and the thorny-rib blackberries. The old Hayman house, now 156 years old, rises brown and sturdy ahead of us. I envision the pumpkin pie baking in the kitchen oven, its scents of cinnamon, ginger, and cloves wafting through the house. That pie, along with whipped cream and sandwiches, will accompany us to Bible Hill, where the food will rest in the fridge in Vilis's lab at the agricultural college until, weary for a while from walking and watching, from hot sun and midway music and cotton candy, from barrel racing, sheep dog trials, and the horse pull contest, we come for supper. Then sandwiches and pie and its whipped-cream partner will re-energize us for more of the Nova Scotia Provincial Exhibition, for car and motocross races, 4-H livestock, team penning, pole bending, and midway lights bright against a clear night sky. Today is fair day.

I glance at Jānis and remember another fair day. Five years ago, when he was one and Dainis was four, Vilis, I, and their Aunt Astra took the boys to watch the demolition derby at the exhibition. By the time it was over—the cars crunched, the air grey with dust and exhaust—darkness was falling, thick and soft on a late summer night. As we left the grandstand to walk out of the fair grounds to the car, the lights of the midway caught the boys' attention—the Ferris Wheel turning, the School Bus swinging up and down, the Merry-Go-Round spinning.

"Can we go?" Dainis pleaded, and so, we took him and Jānis in for a while, and Astra bought them tickets to ride the

Merry-Go-Round. When the carousel stopped, Vilis set Dainis on a white charger and stood beside him while I lifted Jānis onto a brown horse and held him steady as he stared at the lights circling the carousel's floor and roof and bordering other midway rides. The Merry-Go-Round began to move, round and round, the horses rising and falling, the lights swirling and twinkling white-bright as Jāni continued to stare, silent, his mouth open, his eyes rounded, huge, and filled with the enchantment of the midway on a warm summer night. Years later, some of the fair-delight still lingers.

August 24 (20°C, clear, calm, 7:15 p.m.)

I took off the black-and-white coat today, hung it up, and felt relief. My heart tells me that I'm more writer than artist. When I walk, my daydreams are word dreams, not picture dreams.

After supper, my men and I stroll up the back road, the air warm and sweet around us. Jānis decides he wants his bike and runs home to get it, his tufty hair bouncing on the crown of his head. In a few minutes, he rejoins us, and I see, for the first time this year, striped maple seeds hanging at the roadside like clusters of green, winged grapes.

Farther along the road, Vilis notices a carrion beetle, black against the gravel, its four orange rectangular spots like warning signs on its back. The burly beetle crawls slowly, powerfully, like a six-legged tank. Dainis spots a dead frog—flattened, belly up—nearby, and Vilis sets the carrion beetle beside the frog to see what it will do. The miniature living tank crawls under the frog and hoists it upward, only to reveal a flurry of invertebrate life already under the dead amphibian.

Another carrion beetle lunges into motion, and smaller beetles and tiny insects or arachnids whose identity we can't determine in the dim forest light race out from under the frog and over its belly.

The boys and Vilis mark the frog's position with three sticks pointed toward it from different directions. Then we continue our stroll/bike ride as far as Balmoral Road. On our return, we see that the carrion beetles have moved the frog at least eight centimetres from its original position. Where will they take it? Will the menagerie of other decomposers accompany them? We humans think of rotting flesh as stinking and repulsive. To these insects (and possible arachnids), rotting flesh is food, sustenance, the death that gives them life. When they have consumed that rotting flesh, the stink will be gone and the air will be sweet once more. Come, beetles, and feast. You are welcome.

August 28 (11°C, mostly cloudy, SW wind, 6:45 a.m.)

This morning, the risen sun is an orange smear in eastern clouds. A yellow leaf falls from the pincherry tree whose sweetly perfumed flowers of early June made me want to be a bee. Sawdust lies sprinkled on the back road, sign that the miller was here. For five hours yesterday, his portable saw screamed as it peeled slices from our beaver-cut aspen logs. Now we have tidy piles of two-by-fours, two-by-sevens, two-by-tens, one-inch boards, and slab wood, all stinking like urine.

A lone eastern wood-pewee sings in the backroad hemlock woods, its mournful "pee-a-wee" song clear against light wind in the trees. The Nova Scotia dawn chorus has thinned to a whisper. With the breeding season over, migratory songbirds—those that haven't already left us—are feeding, feeding, feeding in preparation for the long flight south. Soon only our resident avian neighbours will remain—the blue jays, with their strident calls; the goldfinches that chirp "pot-a-to-chip" in flight; the chickadees, with their smart black caps; and others.

One morning in the dead of winter, I opened my bedroom curtain and saw, silhouetted against an ultramarine sky, a barred owl perched in the young elm at the edge of our yard. The owl's identity was unmistakeable—the large, oval shape, the rounded head with no protruding ear tufts. Joy swept through me, the joy of receiving an unexpected and delightful gift. How long had the owl perched there motionless and silent like a watchman guarding his treasured homeland? Was it there while I slept? I like to think it was. I like the idea of the owl's keen chocolate-hued eyes scanning our yard and meadows, its talons and silent wings poised for the hunt.

Yet, what of the hapless deer mouse or meadow vole whose body would be crushed by the owl's talons? Too often, I assign value to creatures based on their positive or negative impact on me. I love birds, so I cheer for the barrel owl in one corner of nature's boxing ring, applauding it as it swoops toward the small rodent sprinting for its life in the other corner of the ring. Predators, especially attractive and powerful predators, are charismatic. They draw attention merely by the danger they radiate. The polar bear that killed a seal at Alexandra Fiord shocked me with its predatory strength. Its presence on the ice, with the doomed seal clamped in its jaws, drove home to me the fragility of my existence. And I earlier wrote that I've never seen another single creature exhibit such power as the polar bear that frolicked on a rocky island in frozen-over Alexandra Fiord and then loped over the ice toward the head of the fiord. But what of the slow-moving carrion beetle that hefted up a lifeless frog, a creature whose mass was many times greater than its own? That beetle also was magnificent. It also displayed strength enough to steal breath away. Am I a prejudiced viewer of the earth's biota? I fear I am. My joy on seeing the barred owl, my wonder at that early morning gift, might well have come at the cost of at least one rodent's life.

On Balmoral Road, scats mark the gravel shoulder. We humans who travel this highway don't realize how many other creatures travel it as well. On Matheson Brook Road, the wind speaks louder, its voice like the rush of riffles on a big stream. Change is in the air. Today holds none of yesterday's calm laziness, nor its promise of still heat delivered to the accompaniment of cicada, cricket, and grasshopper orchestra. Yesterday's languorous heat made me slow down, made me listen, made me feel this end of summer as though it were a kiss on my lips and in my hair, a kiss sliding down my shoulders to my thighs, my toes.

Matheson Brook, when I reach it, runs quietly. Brown trout swirl in clusters in the deep pool under the bridge. Up the hill, golden sunflowers in the front flowerbed greet me, as does the reek of fresh-sawed aspen.

Yesterday, I brought my drawings home, the ones that didn't sell, and felt an unexpected sadness that it was all over. The exhibit. The days of focused artistic challenge. My first instinct was to get busy with something, but then I realized that I was mourning, and that it was important to let myself mourn. So, I touched the drawings tenderly and found homes for them on my walls.

Now, in late afternoon, a hot wind blows over Ravenhill. My house is clean, and six jars of dill pickles stand cooling on the kitchen counter. My sons and I are at the brook, where I pick raspberries while Jānis fishes for trout and Dainis constructs a low rock dam and then plays with his cat, Blotchy, on the bridge. This is a time of just being, of enjoying summer's gifts, of anticipating "ultimate summer pie," a banana and coconut cream pie with the tangy surprise of raspberries hidden under whipped cream. Small delights, small promises kept.

August 30 (13°C, overcast with drizzle, N wind, 8 a.m.)

This morning, I'm greeted by a rain-washed world of goldenrod and aster, glistening spiders' webs, and puddles on the back road. It's been a long time since there were puddles. In the forest, wind rushes through the treetops, spattering me with raindrops knocked from hemlocks, birches, and beeches. No pewee sings. Nova Scotia rests on the brink of autumn. At the edge of Balmoral Road, seed heads of pussy-toes bunch into furry-looking clumps. Beside them, bracken ferns raise tattered half-green, half-brown fronds, and the dried stalks of evening primrose are gaunt, faded bones. Today is Saturday, and the road is quiet without the traffic of weekday commuters. A red squirrel chatters from the forest, a junco calls its "tsuk" note in roadside shrubs, and a robin flushes from the ditch. No weekday/weekend distinction exists for these creatures. Every day they must keep at their job of survival.

On Matheson Brook Road I pick up a Molson Canadian beer can and put it in a litter bag I carry. I'll give the can to Dainis and Jānis, who will exchange it for a nickel at the bottle depot. Farther along the road, I collect a paper Pepsi cup, then a Coors Light beer bottle, and then a Labatt Wildcat beer can. I feel like I'm a social anthropologist delving into the inner lives of Balmoral Mills residents. I can see it now—my research paper published in an eminent journal, my work titled, "Drinking habits of rural Nova Scotians based on evidence of discarded containers." Near our mailbox, I collect a Pepsi can and a second Labatt Wildcat can. So far, beer is ahead, four to two.

I turn onto our front road and hear Matheson Brook's peaceful murmur. Raindrops drench roadside grasses. A wasp nest is a wet grey sphere, and a spider web in a hole in the bridge looks like lacework of glistening rain-pearls. The pincherries near the bridge still bear brilliant crimson fruits, but a red maple near the pond is already losing leaves. They fall, hot coral-pink, onto wet, dying green.

Up the hill, I set my litter bag on the porch and then toss raspberry and blueberry cleanings to our chickens, fill their waterer, and open the coop doors. Granny and Fuss-O remain in the coop, glued to two nest boxes. They're incubating eggs. Granny has five, and her granddaughter has four. When I come near, they hunch down and flatten themselves protectively over the eggs, the tufts of feathers on their cheeks puffing out in bristled warning. Fusso squawks loudly, and Granny, softly. I stroke their heads and necks and croon "shoo-shoo" to quiet them before carefully lifting each, in turn, to check for new eggs.

Although the broody hens don't lay new eggs, our other four hens insist on jumping into the nest boxes with Fusso and Granny to lay *their* eggs. The current situation is an improvement. Originally, the nine eggs under Granny and Fusso, each of which is marked with a black spot to distinguish it from newly laid eggs, were in one nest box with both Granny and Fusso incubating them *and* the other hens scrambling on top of the broody hens for their chance to lay eggs in this one alluring nest box. Vilis and I successfully separated Granny and Fusso, assigning each of them a clutch of eggs to incubate. Then Vilis placed an empty nest box, newly filled with dry grass, right beside the broody hens to offer the other layers a choice spot to lay their eggs, but they completely ignore it. In this coop, some magnet pulls all eggs toward one destination. Do the non-broody hens instinctively seek to place their eggs where new life will hatch from them? Who knows what goes on in the mind—or hormonal glands—of a chicken?

I step out of the coop and pull the feed room door closed behind me. The harsh sound of the catch grating into place spooks Blotchy, who skitters away and then returns at my call, romping and playful, her arching back and butting head inviting my caresses. Her fur is soft and warm. She scrambles up the back of my coat, and I laugh, then rise to my feet and watch her leap away. Warm cat. Warm hens. I touch them and they touch me, in that place inside me that needs to hold soft living things.

A week ago, Dainis sat beside a bonfire that lit the darkness. Tears trickled down his cheeks because it was time to go, to leave friends and the delight of a bonfire on a summer's night, to leave the kitten that lay sleeping in his arms, one young hand stroking its fur. When does hardness set in? The masking of open, tender emotion? The holding back of tears that would trickle down the cheeks of one whose heart is torn? Too soon. I hope my sons will never be too grown-up to cry.

September 2 (12°C, overcast, calm, 6:30 a.m.)

This is the kind of day that closes you in. Haze in the air. Quiet songbird voices whispering in the forest. With summer solstice two months in the past and the autumnal equinox fast approaching, the woods are noticeably darker at 6:30 a.m.

A squirrel's chatter is a beacon of bright sound. I step up my pace to evade a swarm of buzzing wasps. Calls of black-capped chickadees announce the perky songbirds' presence, and I imagine these avian forest sprites flitting from branch to branch, sometimes hanging upside down from twigs in their quest for food. Of chickadees' talent for finding nourishment, Aldo Leopold wrote, "In winter, when we are harvesting diseased or dead trees for our fuel wood, the ring of the axe is dinner gong for the chickadee tribe. They hang in the offing waiting for the tree to fall, offering pert commentary on the slowness of our labor. When the tree at last is down, and the wedges begin to open up its contents, the chickadees draw up their white napkins and fall to. Every slab of dead wood is, to them, a treasury of eggs, larvae, and cocoons. For them every ant-tunneled heartwood bulges with milk and honey."[32] For the chickadees whose bright voices I hear this morning, may the backroad woods provide much milk and honey.

Balmoral Road stretches before me like a blank slate, as though it were the day ahead of me. Come evening, what words, actions, and thoughts will be written on my slate? I think of a quote attributed to many sources, including anonymous proverbs, William Penn, Emerson, and Gandhi: "I expect to pass through this world but once. Any good, therefore, that I can do, or any kindness that I can show for any fellow creature, let me do it now. Let me not defer or neglect it, for I shall not pass this way again." We all have a one-way ticket through life. We may never know if our words and actions have a positive long-term impact on the world around us, but we can perform simple acts that show immediate results, like picking up beer cans and potato chip bags. Even better, we could *not discard* those beer cans and potato chip bags. That would make a difference. That would be an act of kindness to the earth.

A truck roars on Highway 311 to the west, but my immediate space is unbroken by man-made sounds. Cedar waxwings trill their high, lilting songs in the trees. Overhead, white-winged crossbills chitter in halting bursts. When I turn onto Matheson Brook Road, I'm struck by a sudden loud hum of insect sound. What is it about the verges of a gravel road that invites insects whereas the verges of a paved road do not?

A frog, squashed by an automobile tire, lies exploded on the gravel, its juices spurted out in all directions like a death star. As I study it, a horde of mosquitoes appears—hence, the loud insect hum. One flies into my ear; others bite my face, arms, and the back of my knees. I step up my pace to try to escape but am engulfed by the mozzies' buzzing drone and stinging bites. I suppose this is fitting, since today my sons and

I will begin our new academic year with a study of—you guessed it—*insects.*

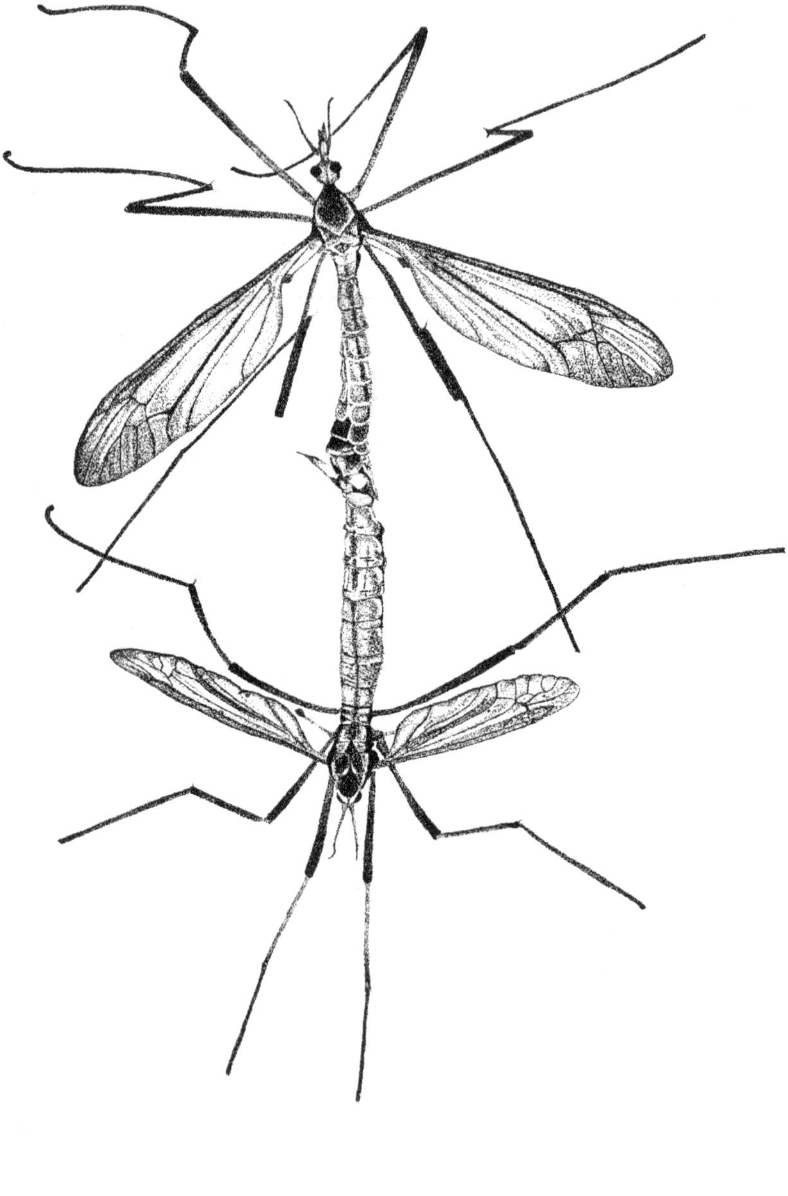

Now it's late afternoon, the first day of our new school year behind us, and the boys and I are walking to the mailbox. As we pass tangled raspberry canes growing near the bridge, I spot a fuzzy caterpillar on a raspberry leaf. The caterpillar, banded with black, white, and yellow, doesn't move. I walk closer. There's something odd about this lepidopteran larva's utter stillness, even when I loom over it. It looks a bit crumpled, and three brown pins seem to be poking into it. Then my wits grow sharper and, as if I was blind before and now granted vision, I see three stink bugs, brown and inconspicuous, behind the caterpillar. The "pins" are the stink bugs' long straight proboscises jabbed into the caterpillar, impaling it, holding it captive while the predatory bugs digest their prey's insides and suck them from its body. For a moment, my breath falters.

Just this morning, Dainis, Jānis, and I read about assassin bugs in an issue of *Ranger Rick* and saw a photo of one of the slender green insects feeding on a caterpillar. In that image, the assassin bug's long proboscis impaled its prey in the same way that the three stink bugs' long proboscises impale this caterpillar on the raspberry leaf. The article was informative, and the photo, intriguing, but both pale in comparison with the brutal immediacy of this life-and-death drama in front of me. I crave beauty and long to create beauty, so I look for beauty in nature and relish the wonder it affords. An unexpected and beautiful sight stirs the heart like no other. The scene before me now is a vastly different kind of unexpected—a shocking unexpected. As with the polar bear clamping a seal in its jaws, nature reveals its hard underbelly. Death lurks behind a raspberry leaf. Survival belongs to the seal that detected a polar bear creeping over the ice toward it, and then slid into the fiord, but not to the seal a polar bear clamped in its jaws, and not to this thrice-punctured caterpillar.

I show Dainis and Jānis the caterpillar and stink bugs. They're thrilled, fascinated, excitement emanating from their voices as they lean in for a closer look. At times like this, when book learning flowers into reality before our eyes, we want homeschooling to go on and on.

September 3 (16°C, overcast, strong S wind, 6:30 a.m.)

The wind is warm and strong, like a good man's arms. A songbird flits along the back road, but the early morning light is so dim I can't identify it. When I reach Balmoral Road, clouds billow tall above me, violet-blue and moving fast. These days, with daylight grown shorter and avian heralds of dawn muted, it's harder to rise at 6 a.m. to get out into the world. That said, this road is the same solid surface beneath my feet, stretching my legs and lungs, my mind and eyes.

If I could, I would give a bit of advice to every resident of Balmoral Mills, of Nova Scotia, of Canada, of Earth. Go for a walk today, or a run, a climb, a ski, a paddle. Or just get outdoors. Look. Listen. Feel. Rain might plaster your clothes to your body. Wind might whip snow into your eyes and burn your hands. Go anyway. Do it. You might see a sleek dark mink by your brook, hear the rough "ho-ho-ho-hoarrrr" of a barred owl, or feel winter's chilled breath on your face as you gaze at the Big Dipper, Orion, or Cassiopeia. *Go. Do it.* And when you do, the wild spirit of life within you will wake, laugh in your ear, and whisper, "Now *this* is living!"

Once during the first winter that Vilis and I lived in the log cabin on the shore of Kluane Lake in the Yukon, I accompanied a young researcher, Nick, on a hooting survey. The month was March, still fiercely cold. My flower box of pansies on our cabin wall was long asleep, and I was seven months pregnant with Dainis, whose shrill baby cries would halt my writing frenzies during his naptimes. Nick and I drove the Alaska Highway through calm evening darkness, stopping to listen for owls hooting near each of nine experimental snowshoe-hare-trapping grids. At each stop, Nick played a tape of owl calls starting with small saw-whet and boreal owls, since they wouldn't call after a great horned owl had called (the big guys eat the little guys). After each owl's call played, we listened for responses. Above us, green ribbons of northern lights rippled across blue-black sky studded with stars in an infinite chandelier hung in the heavens. A few concerned motorists stopped to ask if we were okay, having seen my big belly pushing against my down jacket. In the Yukon, people look out for one another.

That evening, Nick and I heard eleven great horned owls hooting, several of which at first called faintly from a distance and then moved closer in response to the challenge of the tape. At our last stop, we opened the truck doors and heard wild clear voices echoing like moon song in the still night air. Not owls. *Wolves.* The pack howled from a steep-sided hill that plunged down to Sulphur Lake, a hill silhouetted against ultramarine sky lit with the Aurora Borealis. I grabbed that wolf song and hid it inside me, and the wild spirit of life laughed and whispered in my ear, "Yes, *this* is living!"

So, go beyond walls and roofs and see and hear what awaits you.

September 5 (6°C, clear, calm, 6 a.m.)

I hesitate to open the patio door. A doe and two fawns are feeding on fallen apples under a front yard tree, their bodies slim and brown, their ears always moving, always testing. I slide the door open as quietly as I can and step onto the porch. Instantly, the three deer cup their ears in my direction, their eyes fixed on me. As I descend the porch steps, the trio bounds away gracefully, tails flashing brilliant white against lush dawn foliage. Our three visitors are off to where morning will take them, and so am I, with the air cool and fresh around me. Gone are the warm nights and wet days of the past week. Today will bring sunshine.

As I walk the back road, my mind is filled with nothing but an image of breakthrough. The scene plays over and over, as though nudging me toward revelation. This morning, after Vilis and I rose from our bed, he spotted the doe from our upstairs bedroom window. Then something shifted—a faint movement in the dawn—and a fawn was at her side, sturdy and graceful. We watched the two deer in open view and saw nothing else until again something shifted in the dawn and a second fawn appeared right there on the lawn, flowing from invisible stillness into eye-catching motion.

I think about the caterpillar impaled by the three stink bugs' proboscises, about how I didn't see the stink bugs at first, just like Vilis and I didn't see the fawns at first. My view was too narrow, too restricted. I wasn't looking beyond the caterpillar with its bright bands of white, yellow, and black. It was only when my perception switched to a broader, more encompassing view that I realized that the thin brown "pins" were stink bug proboscises. It was only when something in my perception shifted and changed that I saw what was there all along. How often are there occasions in our personal lives when something shifts, something changes, and we suddenly see what was there all along? Love? Envy? Betrayal?

On Balmoral Road, the scent of sheared spruce hangs fragrant in the air, drawing me from my contemplation. Spruce, birch, and alder saplings spear the ditch sky, torn and mangled, parts of their limbs spewed onto the pavement by a brush cutter. The ditch isn't the only shorn bit of landscape. A clear-cut hillside looms on the south side of Balmoral Road. Last year, in the dark of December, loggers cut day and night, the noise of feller-bunchers working a kilometre away from our house so loud it woke me from sleep. Now, all that remains is a small stand of pines on a hilltop and a copse of mixed woodland. The rest of the forest is gone, clear-cut in one fell swoop.

The tendency of animals to stay in or return to a certain place is known as philopatry. Many birds show a high degree of philopatry. This past spring, migratory birds that had previously nested in the woods on the now-clear-cut slope would have returned from their wintering locales and found the forest completely devastated except for the hilltop pines and the lone stand of mixed woodland. Instead of their familiar nesting habitat, those birds would have arrived to find stumps rather than trees, and branches strewn over the ground. Where did all those birds go? Did the loss of their breeding habitat create increased competition for nesting sites in nearby forest stands? Did some birds fail to breed? What are the long-term consequences of the fragmentation of breeding habitat (due to clear-cutting and other disturbances) for our migratory songbirds?

We must remember that we don't walk this earth alone. Anthropologist Bernard Campbell wrote, "There's no escape from our interdependence with nature; we're woven into the closest relationship with the Earth, the sea, the air, the seasons, the animals and all the fruits of the Earth. What affects one affects all—we're part of a greater whole—the body of the planet. We must respect, preserve, and love its manifold expressions if we hope to survive."[33]

Clear-cutting a forest does not respect the body of our planet. In 1978, I wrote a university essay that discussed the deleterious effects of clear-cut logging on trout and salmon streams. My essay was based on research that was at least ten years old; thus, people have known for at least thirty years that clear-cut logging increases runoff, depletes long-term moisture reserves, increases soil erosion, raises river temperatures, and smothers trout and salmon spawning beds with silt. And still the clear-cutting continues. I see no conscience in it, only greed.

As I walk on, past the clear-cut, my mind returns to the concept of breakthroughs. Yesterday afternoon, as part of our insect study, Dainis, Jānis, and I took dip-nets, trays, and a bucket down to the pond, where we dipped, swished, and swept for aquatic insects. We netted diving beetles, dragonfly nymphs, water boatmen, marsh treaders, and saw damselflies perched blue and slender on lime-green grasses. We netted other creatures, too: tiny clams, minnows, green frogs. One of the frogs was so huge the boys thought it had to be a bullfrog. We got stuck in the mud at the pond's edge, the boys shouting with mock-dread when they lost their boots and stepped sock-foot into cold squish-stickiness. Their feet and hands got muddy, and soon they were smearing mud up and down their arms and legs, with huge grins on their faces, their goofy laughter filling the air.

Afterward, we collected our mail and, on the way, found eastern red-backed salamanders, a pill bug, and the sleek pupa of an unidentified insect. Treasures surrounded us, there for the finding—a giant grasshopper in roadside grasses, a blue flea beetle on alder leaves, and up the hill, a pale green lacewing on our screen door. The boys hosed themselves off and showered. They want to go down to the pond again soon. They want to take the canoe and dip for tadpoles and red-spotted newts and that really big frog they think *must* be a bullfrog. I want to go, too—to catch their excitement, hear their laughter, find more treasures, and discover whether I also can shift and change and, in doing so, see things I've never seen before.

September 9

It's evening, warm and misty. I'm slicing onions for bread and butter pickles when memory again straps me into a time machine. In the same way that walking with Charcoal or picking up a cigarette package triggered vivid memories, the pungent scent of onions transports me to a different time and place. I'm back in Viking, Alberta, in Grade 9, at a school dance. I'm wearing a soft, rusty-orange sweater and dancing a slow dance with my date in the gym. He draws in a long sniff and laughs gently, saying, "You smell like onions." I giggle and tell him that my mother was making mustard and onion pickles, hence the sweater's fragrance.

I telephoned Mom earlier this evening. She told me that my brothers are combining canola on the hill kitty-corner to the southwest corner of the home farm. Then she looked up from the phone and saw they had moved, were directly across the gravel road from her trailer, the combine's lights, brilliant in the early dusk.

We are where we came from, always. At least some part of us is. Tonight, all of me is, and there are no hard memories, only soft laughter, a combine's lights in the dusk, and the smell of onions, rich and pungent, in my mother's kitchen.

September 10

I'm rushing—rushing to feed the chickens and to bottle my bread and butter pickles, rushing to grade my sons' schoolwork and to get the boys into the car to attend swimming lessons. Still, I somehow make time to bake a poppyseed chiffon cake while the pickles boil in the canner, and to glaze the cake with tart lemon icing while the boys shuck corn for supper.

Exactly six years ago, Vilis and I signed the papers to purchase this property we call Ravenhill. Dainis was three then, and Jānis, only four months old. At supper tonight, my sons and I (Vilis is away at a conference) feast on corn grown on our land, after which I raise a forkful of cake in a toast. "Happy anniversary…to this place."

"To the *best* place," Jānis corrects.

"To the best place in the world," I amend.

"To the *one* best place."

"Yes, to the one best place."

I don't rush while driving to Pictou for the swimming lessons, or while watching the boys work hard in the pool, or on the way home. For a bedtime snack we eat more cake—tall narrow slices of celebration. Then we sleep like we did on that first night, with our land wrapped around us and the music of crickets in our ears.

This old overgrown farm on the shores of Matheson Brook is a gift. It's the red box treasured by the sky people, although instead of holding one rooster it contains entire communities of animals and plants. When the box is opened, life bursts forth—crowing, hunting, escaping, mating, existing within the vast panoply of Creation. And the sun rises, bringing light to the world, casting its radiance onto this place—our place—that casts its own radiance into our lives.

September 12 (22°C, light cloud, brisk S wind, 4 p.m.)

It's late afternoon, and I'm on the back road, walking strong, stepping into the remainder of the day. A sudden slithering movement on the gravel catches my attention. The sinuating motion belongs to a slender snake, golden tan in colour, with three yellow spots like faint beads on its neck. Since no poisonous snakes naturally occur in Nova Scotia, I pick up the pretty ophidian to check its belly colour, which is orange. My captive is a red-bellied snake. Agitated at being imprisoned, the snake whips my hand with its tail. As I release my prisoner and watch it slither away, energy whips through me as the snake's tail whipped my hand.

I learned a lesson today. When an old emotional wound begins to open and threatens to spill your guts on the ground, turn away. Do something that makes you feel good and strong, that reaffirms to your innermost being that no matter what anyone says, does, or thinks, you are a person living a life of worth, and within you a spark exists that can in some way make this world a better place. A corollary to the above lesson is to do in life the things you love, for it is those pursuits that will make you feel strong and valuable. This morning, I phoned the woman with whom I'd once shared a vibrant friendship, a friendship I destroyed with one tactless, hurtful comment. I thought time might have soothed her pain, but her brittle tone of voice and brief response conveyed no desire to speak with me. After the call I had to make myself stop doing agitated chores and mentally turn away. I took a hot shower and focused on the feel of clean skin and long hair. Then I made myself write, and in the writing again found peace and confidence.

My strides eat up Balmoral Road, then swallow Matheson Brook Road. Mud kicked up by vehicles has spattered asters, fireweed, and our mailbox, leaving them grey and marred. The brook is clean, though, and runs clear beside our front road. Near the bridge, I pluck blackberries almost hidden by leaves. They're earth and sky in my mouth, sun and soil on my lips, like wine—autumn wine.

September 15 (12°C, mostly clear, SW wind, 7:45 a.m.)

A raven calls above the back road, its voice low and guttural. My children cycle after me, asking if they may open the coop door so they can shoo out the hens and roosters. I say yes, and they ride home. The day is filled with promise, as much promise as rests in the shrill cheeps of one newly hatched chick, the chick my sons want to hear and see without the noise and distraction of excess fowl in the coop.

The raven flies over me, whipping air, a long narrow gap in its right wing indicating where a primary feather is missing. Blue jays mutter, and a red squirrel chatters its territorial "chirr." The wind sighs, and so do I. I'm free—free from motherhood and teaching. Vilis is home, returned from a conference a continent away, and he's offered to teach the boys two mornings a week, shifting his work schedule on those days to afternoons and evenings. Today is the first of those days.

I didn't know until this moment how loud all-terrain vehicles can be. A four-wheeler makes the turn from Matheson Brook Road onto Balmoral Road, its engine's rough roar hammering the sky. Cedar waxwings flush from roadside trees, their wings fluttering like leaves against the blue bowl above me. It's good to walk again. I had few chances while Vilis was away,

On Matheson Brook Road, three deer—two does and a fawn—step from the ditch onto the gravel and freeze into instant wariness. They've seen me. I continue walking toward them, and the deer break into sudden motion, trotting swiftly across the gravel road, then disappearing in roadside shrubs. Around a curve on Matheson Brook Road, another doe crosses the gravel, her brown coat rich and warm in the morning sunlight. Where are these deer going? To rest in the forest and digest their dawn meal?

Nova Scotia is a good place for deer. I see them more frequently here than anywhere else I've lived in Canada, although the greatest numbers I've seen were in southwestern Manitoba, where Vilis did his skunk research. There, in late autumn, herds of sixty or seventy deer fed in hayfields on Turtle Mountain's rolling hillsides. The sight of those herds made me want to travel to Africa to see even larger herds of ungulates—wildebeests, zebras, gazelles—on the veld, and to explore the ecosystems in which they live. That, however, is a dream for another time.

An American kestrel flies overhead, its pointed wings beating rapidly. Low morning sunshine highlights the bird of prey's rufous breast. Is it hunting? Kestrels, which are North America's smallest falcons, are thought to have a secret weapon when it comes to locating small rodents, particularly meadow voles. Like many insects but unlike humans, kestrels can see ultraviolet light. When meadow voles move about, they leave urine trails that reflect ultraviolet light, so those trails are visible to kestrels. I try to imagine what the earth's surface looks like to a kestrel. My mind envisions hot blue streaks and squiggles searing the ground surface, some of which lead straight to potentially luckless voles.

The hill to our yard seems long and steep this morning, so I push harder to conquer it. On reaching our porch, I remove my sneakers and set them in the sun to dry from the dew.

Indoors, Vilis and the boys are reading about termites in an issue of *Ranger Rick*. Dainis and Jānis show me a photograph of a termite queen, her abdomen huge and distended. Excitement flowers in their voices as they tell me that the termite king and queen can live for *fifteen years*, and the queen can lay *one egg every second.* I take out a calculator and do the math. One queen could lay 473 million eggs in her lifetime. *One queen.* It's a wonder that termites don't rule the earth.

Of course, termites are but one group of insects, and it could be said that insects really *do* rule the earth, since they comprise eighty percent of all living animal species. Beetles alone make up more than a third of all insects and, thus, are the biggest group of animal species in the world.[34] I think again of the powerful carrion beetle hoisting up a dead frog, and of the shiny blue alder flea beetles crawling over lacy, skeletonized alder leaves. In the grand scheme of life here at Ravenhill, these are some of our small representatives of a huge and mighty insect order, the Coleoptera, without which our world would not be the same.

In late morning, mid-September heat wraps me in woolly somnolence. I'm picking blackberries in the patch below the pine windbreak east of the house, where weeks ago a deer bolting into motion awakened my senses to a new day and fulfilled Thoreau's recommendation of an early-morning walk. Amid the thorny brambles, I'm surrounded by other of Ravenhill's resident insects—slow-moving wasps that cling to ripe berries, their bodies, vivid yellow-and-black curves warning my fingers away. My vespid companions stagger over the lush blackberries as though drunk on their sweetness, making it easy to avoid them. As I pluck rain-gorged fruits, their thick, sticky juice spreads purple stains over my fingers.

During June and July, Ravenhill's soil begged for rain, yet within the past month, we've received so much precipitation that tomatoes split on the vine and blackberries mould on the cane. With my mouth stained as purple as my fingers, I move drowsily through the brambles, as drunk on the berries' sweetness as the wasps.

Warm night, big moon, west wind. It's a lover's night and a harvest moon. The harvesting, I did earlier—blackberries this morning and potatoes and onions this afternoon. The boys helped with the spuds. Dainis dug, and Jānis picked, and then while I dug, Dainis picked, and Jānis pushed the wheelbarrow up and down the rows of soft soil, his load growing heavier and heavier. At the end of the last row, Jāni finally handed over the wheelbarrow, two-thirds full, to me. I lifted the handles, felt the load's weight, and marvelled at my younger son's strength and determination.

But perhaps I should have known. After all, this is the boy who at age three essentially ditched his finger-sucking habit the day our dentist told us the sucking could lead to problems with dentition and speech. That first evening, as Jāni wailed, "But I *can't* stop!" in Vilis's arms, I knew what he was thinking—that he'd always sucked. In fact, he'd sucked even before he was born. An ultrasound image clearly showed our baby's hand held in his mouth, and after his birth, he had a sucking blister on the inside of his lips.

At the promise of a treat *only for him* if he didn't suck, Jāni rallied and stopped crying. We tucked him in bed, put his hands down under the blankets, and pulled Favie, the baby blanket he dragged everywhere with him and without which he couldn't fall asleep, up against his cheek. When I checked on him later, there was no finger in his mouth. In the morning Jāni said he'd wakened three times during the night, had realized he was sucking his finger, and had taken it out of his mouth. He chose a marshmallow for his treat.

Later that day, when Jāni was fighting the urge to lift that dear finger to his mouth, I started to wrap Favie around him so she could soothe him. (Favie, short for Favourite, was apparently feminine.) "No," he told me. "I don't want her. She makes me want to suck." A whole year passed before my little one snuggled with Favie again, her satiny edge against his cheek, the finger he'd once pressed into his mouth now pleating, pleating her softness.

Vilis drives his old red truck into the yard, and he and I go strolling in the night. (We haven't abandoned the boys. Vilis's sister, Astra, is reading in the room next to theirs). We walk through meadows of silver grass and into the backroad forest, where puddles glisten in the moonlight. Beyond the puddles, shadows loom like black walls.

Having spent many nights following skunks in Manitoba, Vilis is perfectly at home in the ghostly ambience, but I'm not. This road, so familiar in daylight, is a wraith leading me on in a night dance of uncertain steps. I cling to Vilis's hand as we feel our way through an abyss of darkness in the Cathedral and then into clearings where moonlight paints white pools on the road and casts stark elongated body-shadows behind us. At the top of the back road, Balmoral Road is a blue-grey ribbon stretching into blue-grey eternity in blue-grey silence, all colour and vibrancy seemingly lost from the world.

My steps quicken as we turn and retrace our steps through the white pools and black walls, past the glistening puddles and silver meadows, home to where golden light spills from an old house under a harvest moon.

September 17 (6°C, partly cloudy, NW wind, 6:45 a.m.)

Goldenrods have lost their gold, their grey heads spewing tufted white seeds in the meadows south of our house. A blue jay's strident call rings out, and chickadees murmur. These are winter birds. More precisely, they're year-round residents.

We narrowly escaped frost last night. At 10 p.m., our thermometer on the porch read only three degrees above freezing. Vilis and I carried sheets and tarpaulins to the garden and covered squashes, pumpkins, tomatoes, and basil. September is a time of guarding, harvesting, and processing, processing, processing—pickles and pesto, dried apples and frozen tomatoes. These are summer's tastes for winter's table.

A gentle wind rustles leaves in the backroad forest. Sparrows whisper, and cedar waxwings trill. On Balmoral Road, sunshine washes over me, and I lift my face to the sky. How I love this world! Its plants, animals, land, and sky are as much a part of me as life itself. In his book of Psalms, King David wrote: "Let the heavens rejoice, let the earth be glad; let the sea resound, and all that's in it. Let the fields be jubilant, and everything in them; let all the trees of the forest sing for joy."[35] This morning I'm enveloped in earth song, buoyed on the wings of Creation. For me, there is no greater boon to my spirit than walking among the beauties of the earth.

On Matheson Brook Road, loud tapping resonates through the air, and I spot a pileated woodpecker high in a dead elm near the brook. This largest of Canadian woodpeckers is the size of a crow, and its folded wings are as black as a crow's but the crest atop its head is a red flash in the sun.

Once when I was running, I spotted a pileated woodpecker hammering into a spruce tree beside this road. It clung to the bark with its zygodactylous feet—two toes pointing forward; two toes pointing backward—and swung its beak like a man swings an axe, from one side and then the other. I stood transfixed as wood chips exploded from the tree. From book learning and that Ornithology class I took two decades ago, I knew that woodpeckers' skulls are reinforced by thick plates of spongy bone to absorb the shock of the birds slamming their beaks into trees. Yet, I never truly appreciated why that cushioning is necessary until I watched that pileated attacking the spruce. I had no idea that a bird could swing an axe like that.

3

Golden Fire

September 23 (6°C, overcast, S wind, 6:15 a.m.)

Dawn brings a quarter moon above me, wind in my face, and a pink sky in the east. Chickadees greet me at the woods, and a robin mutters from within the forest. Autumn has arrived, the days now shorter than the nights. Temperatures drop to single digits each evening as the sun dips toward dusk. Yesterday morning the mercury hovered barely above freezing, yet our garden escaped frost again. I braided onions and picked ripe tomatoes, one of which measured an astonishing fifteen centimetres in diameter and weighed a kilogram and a half. When I first spotted its orange-red sides, I thought it was a small pumpkin.

A dump truck passes me on Balmoral Road, and for a few seconds I breathe exhaust. Then the road is mine again, and my only company is the wind, clouds, and shadowed trees. This is time alone, time to look inside, time to observe and wonder, time to develop "holy curiosity."

A crow calls far away, then closer, and I see it black against a grey-blue sky, its wings beating, its body sailing with the wind. The sight reminds me of when Dainis was five and wanted so much to fly. Day after day, he built airplane models using a hot glue gun and heavy paper. His first aircraft were thick-bodied and clumsy; the later ones, slim and streamlined. He saw them in his mind, made them with his hands, hung them in his room. And he searched for wings. In the meadows, in the forest, each time we went walking he looked for wings to wear so he, too, could fly. Vilis and I never told him he wouldn't find any. We couldn't bear to destroy his dream. So, Dainis continued to search and continued to build, his imagination soaring, winging, flying with the models that raced through the skies of his bedroom—biplanes and gliders, passenger jets and aerodynamically trim jet fighters.

A while ago, I mentioned to Dainis how he used to look for wings. He smiled gently, his mouth curving in fond reminiscence. At that moment I saw him in my mind as I see him now in my mind—a small boy in dark blue parka and snow pants, with winter-gold grasses around him, his brow furrowed, his eyes far away, yearning. "Mama," he asks, "will you help me find my wings?"

Matheson Brook whispers. The front-road hill wears autumn's green and gold. I let the chickens out of the coop, fill their waterer, and hope that somewhere, sometime today my sons will have a chance to fly.

September 24 (1°C, cloud/clear, strong SW wind, 7:15 a.m.)

This morning, the wind carries a hint of snow in its touch and ice in its breath—the promise of winter. Several days this past week have been unseasonably cold, with daytime highs 10°C below normal seasonal values. Already, we're burning split logs to chase away morning chill, with the boys practising their fire-lighting skills. Six cords of firewood delivered yesterday are a welcome sight in our yard.

Faint bird sounds dent the backroad silence. White tail edges flash, revealing the presence of dark-eyed juncos, but their movements and noises are secretive and inconspicuous. No males trill boldly from treetops now. Autumn is a time of feeding, resting, and preparing to go—or preparing to stay. Juncos do either in northern Nova Scotia, depending on weather conditions. One mild winter, we had juncos pecking up sunflower seeds from under our feeder for much of the winter; another year, the petite grey-and-white songbirds didn't reappear until March.

On Balmoral Road, wind flips up skirts of roadside maples, exposing leaves' silver-green undersides to all who pass. The leaves' upper surfaces are faded green, some edged with yellow, some spotted with rust and burgundy. All over Balmoral Mills, all over the Cobequid Mountains, all over Nova Scotia and eastern North America, maples are breaking down the green photosynthetic pigment, chlorophyll, in their leaves and reabsorbing its components. As they do so, their yellow, orange, and red pigments—xanthophylls, carotenes, anthocyanins—that were masked by chlorophyll are revealed in vivid glory. Death is beautiful in a maple leaf.

A bitter northwest wind blows clouds to the southern horizon, leaving a pale blue sheet in its wake. My running clothes are thin, so I curl my fingers inside my mittens and tuck my thumbs into my fists. Uncomfortable but invigorating, the cold's stinging immediacy assures me that I'm truly alive.

On the front road, trees drip water onto my toque, and sodden grass drenches my socks and sneakers. During some of my runs, I bask in light and warmth; during others, I endure cold and wet. All these experiences are true and real, all are my feet treading through time as I trace my life path onto the surface of the moment, onto a here and now that is no more than a brief flash within the vastness of eternity. Yet, that flash is *my* flash. My life is a moment in time that will never come again. I pant up the hill, with my footsteps on gravel grating a message: *Carpe diem. Carpe diem. Carpe diem.*

Up the hill, I step into the house, where I'm greeted by the enticing aroma of coffee and by warmth radiating from our wood stove. "Magi," Dainis told me importantly yesterday, "wood warms you twice, you know. Once when you split it and once when you burn it." His eyes shone with the satisfaction of a fire crackling merrily, a fire he had built. Today it's Jānis's fire that flames behind the stove's glass door. My sons are growing up.

September 26 (7°C, overcast with drizzle, S wind, 6:45 a.m.)

Morning is dull and drizzly, the only sound, wind racing through grass and trees. Then a chain saw rips the air—Vilis cutting our newly acquired eight-foot lengths of firewood into stove-size pieces.

The night before last, we had frost. It blackened squash and pumpkin leaves but was, nonetheless, a light frost. I knew it was coming, for the temperature plummeted to 6°C in late afternoon as a north wind chased dark clouds from the sky, leaving expanses of clear blue. So, I harvested the last of my peas and beans, pulled baking beans and hung them for drying, collected Sweet Dumpling squashes and rain-gorged cucumbers and tomatoes, then picked blackberries, racing against nightfall to bring in the harvest.

Now, the backroad woods are gloomy around me, the hemlocks, tall grey statues. Mild air means no stinging hands, and the drizzle has turned to surprisingly warm rain. It reminds me of my family's outing a month ago when we searched for lost aspen logs and then walked to the Amphitheatre. At that time, the colours of rain-washed grasses and trees were rich and bright around us. This morning in the forest, colours are still too muted by dawn to rush against my eyes.

Beside Balmoral Road, wind speaks in trembling aspens, telling a rustling story of where it's going and where it's been. A truck passes me on the pavement, the water spraying up from its wheels silver against the pickup's wet black paint.

On Matheson Brook Road, I hear a raven, faint, far away, high. I step to the edge of the road as a school bus passes, throwing exhaust into my face. My footprints from previous days lie before me, still visible in the gravel, pulling me onward, as though each time I walk or run this loop, I add a few new bricks to the house of my existence.

Life is all about building on what we already have. We lay a foundation in our childhood, add walls in our teens, and a roof as young adults. We install windows and doors as we pursue chosen careers or vocations and fill the dwelling with the furniture of our deepest desires. If we build well, on rock rather than sand, the house of our existence will shelter hope for the future, both ours and that of the earth, and will offer sanctuary for dreams that might otherwise be lost.

I believe that imagination, the seeing in the mind of what is possible, of what *could* be possible, is the greatest driving force for good in this world. Jonas Salk, whose name is synonymous with the polio vaccine he developed, wrote, "Hope lies in dreams, in imagination, and in the courage of those who dare to make dreams into reality."[36] I hope I have the courage to make my dreams into reality—dreams of revitalizing our home, living in harmony with nature, teaching my sons well, discovering more of who I am through exploring forest and meadow.

Robert Bateman, world-renowned for his nature paintings, said, "I can't conceive of anything being more varied and rich and handsome than the planet Earth. And its crowning beauty is the natural world. Whether it's here in Ontario or in the Antarctic or in Africa, I want to soak it up, to understand it as well as I can, and to absorb it. And then I'd like to put it together and express it in my paintings. This is the way I want to dedicate my work."[37] My dream echoes his, although I paint images with words rather than pigments.

On the front road, a chickadee whispers in shrubs. Beside me, Matheson Brook runs so quietly I can barely hear it. Fierce autumn rains—the fallout of hurricanes rolling north from the Caribbean—haven't yet begun during what I facetiously refer to as monsoon season. Once they do, this peaceful stream will become a raging torrent.

At the bridge, the brook speaks louder as it trickles over rocks downstream from the pool. Two years ago, on a mild January day, my sons and I did an experiment here. On our way home from collecting mail, we stopped at the bridge to toss sticks and snow into the rushing water. Dainis was thrilled when a stick he threw in sped downstream. Jānis was disappointed when an ice-ball he threw in bobbed next to the bank. Dainis's stick was moving fast, so why didn't the ice-ball go anywhere? I explained that the stick was caught in the brook's current, and the ice-ball wasn't. "But where *is* the current?" Jāni asked. I told him we'd find out.

Up the hill we went. Out came the hole-punch and sheets of construction paper. We punched and punched and punched, creating small mountains of brightly coloured paper rounds. We put the paper bits into three jars, walked back down the hill, spaced ourselves on the bridge, and emptied our jars over the side of the bridge, tossing the paper like confetti onto the water. The resulting pattern formed by the paper bits was like a V-formation of white, yellow, orange, blue, and pink geese flying on a dark green sky. Fast-moving water in the brook's centre grabbed some of the paper bits, rolled them underwater, and shot them downstream. Mere centimetres to the side of this fast water, other paper rounds followed slow eddies that arced toward the stream's bank. We truly saw where the current was, and where a small boy should throw an ice-ball if he wants it to go *fast.*

Up the hill, Hubbard squashes bulge and curl on the porch. The tang of drying apples spices the kitchen. I water chickens, feed cats, and admire black-eyed-susans, spider plants, and pansies still blooming next to the porch. When all else is fading, these three flowers continue to splash bright colour against cool autumn air and the green of a lawn coaxed from summer desiccation to autumn luxuriance by the past month's frequent rains. When will these blossoms fade and die back, leaving my eyes to seek the stark, barren beauty of winter? Too soon.

On the heels of this thought, I envision sparkling snow diamonds and recall how Dainis taught me to see flashes of pink, turquoise, and gold in the miniature ice prisms. What else will I come to see through my children's eyes? What else will I help them see through their own?

September 27 (8°C, cloud/clear, strong S wind, 11:30 a.m.)

The wind is in my face, and I'm glad that it's strong, cold, and has wildness in it. I'm doing too many things and none of them well. So many mornings, I promise myself a peaceful day, only to end up running—running against the clock, the season, the bank. White clouds scud across blue sky above me. Warm sunshine caresses my shoulders. Fresh deer tracks on the back road are paired pointed peaks trailing across a map. I focus on feeling only the sun's warmth, on hearing only the wind's rush and the musical chirps of crickets. These are the things that keep me sane these days—these and the simple task of putting one foot in front of the other on gravel, on pavement, on gravel.

Matheson Brook speaks to me, saying, "Listen to my song," and so I sit on the old rotten bridge and look for trout. The brookies are hard to spot at first, but then I see tails flick and discern slim torpedo-shaped bodies, their backs mottled with grey and brown. All are small, less than fifteen or twenty centimeters long. I stand and peer into the deeper shaded pool on the other side of the bridge to check for larger trout, of which there are none. Then I walk on. The grasses on the hill are losing their hair, their spiked heads balding in the wind.

In the yard, Dainis and Jānis parachute off a ladder leaning against the porch roof. They've tied strings to blanket corners. I suggest that sheets would be lighter and help Jānis tie strings to a sheet's corners after he and Dainis embrace my suggestion. The new parachutes don't work all that well either, but the boys have fun anyway. Today, I envy their freedom and laughter.

September 28 (-1°C, clear, W breeze, 7:40 a.m.)

Frost whitens our lawn and meadows. No cloud mars the sky. Another raccoon visited our coop, this one digging under the front door. Vilis and I think it's the same coon that has visited our hen house on and off for the past month, digging under the metal sheeting or ripping apart board and batten by the back door—yet we haven't heard any screams in the night. This raccoon dines on laying mash, leaves seedy scats under the roost, and ignores the chickens.

A red squirrel chatters from high in a hemlock in the backroad woods. A ruffed grouse scurries across Balmoral Road, its steps jerky, its head thrust forward like the prow of a ship. Sunlight spills over the horizon, the long-angled early-morning rays flooding roadside tree leaves, lending them the richness of emeralds, bronze, and gold. Now I hear the voices of a late warbler, a goldfinch, and a robin. Are they, like me, welcoming the sun?

On Matheson Brook Road, silvery smears of spit-like mucous mark the gravel. I bend to look closely and note a crushed snail shell. A slug creeps nearby, its slime trail silver in the morning light. Both creatures are molluscs, one with a shell; one, without.

During my second year of university studies, I opened Robert D. Barnes's weighty *Invertebrate Zoology* and fell in love. I read about molluscs, such as this crushed snail and slug here on Matheson Brook Road, as well as about cnidarians, annelids, crustaceans, echinoderms, insects, and more. All were invertebrates, yet so different from each other, so filled with their own spineless character. I learned that earthworms possess both male and female reproductive organs, that starfish move by pumping water through canals fringed with sucker-tipped tube feet, that barnacles suspend themselves in a swimming pool of their own making and kick cirri-lined legs to capture food. Every page brought new wonders and heady terminology.

In a subsequent year, I took a hands-on course about Alberta's aquatic invertebrates. Along with other students, and on my own, I sampled ponds, streams, and lakes, gathering vials of near-microscopic copepods, netting mayflies and mosquito larvae, scooping up a freshwater crayfish, and dipping up caddis fly larvae in their sand-grain, leafy, or twiggy houses. My wits sharpened. I learned to key out stonefly nymphs by checking to see if they had gills on their abdominal segments,[38] and mayfly nymphs, by checking their mandibles for an "external projecting tusk visible dorsally."[39] After one fieldtrip, I dreamed of a great tangled mass of stonefly nymphs, their chiton-armored bodies like those of imagined fearsome aliens invading the earth. Three weeks ago, when my sons and I sampled our pond for insects, I dug out my old textbooks and fell in love with invertebrates all over again.

On the front road, our old bridge is slippery with frost and stencilled with raccoon tracks. Mist hangs over the pond and among shrubs and trees. I scale the hill and from the kitchen watch the mist rise and billow, silver-white in the sun and then—gone.

September 29 (mild, overcast with rain, strong wind, late afternoon)

My feet are dirty, and I stink of exhaust. Blisters sting my hands, my arms ache, and matching bruises mar my upper thighs where the tiller's handles knocked against me while I fought its bucking crawl. That said, my garden is tilled, and not only tilled, but also fed with compost and sown with buckwheat as a late green-manure crop. Drizzle fell as I scattered the first seeds, and I got drenched when I ran to put away tools. Still, as when planting the garden in June, I beat the storm.

Two days ago, there was no storm. Vilis said the weather was perfect for canoeing. I said it was perfect for clearing my garden. And, thus, we clashed—my husband, the hunter-gatherer, and I, the gardener. Ultimately, I gave in since, as Vilis pointed out, my garden would still be there the next day. We drove to Horton's Point near Wallace and explored the shore, picked rose hips, and canoed across Wallace Bay to check out Highlander cattle grazing in a pasture on the opposite shore. The day truly was perfect for canoeing, yet I was miserable. I wanted only to be home, with the touch of garden fruit on my hands.

So, yesterday I cleared my garden—picked the last tomatoes, squashes, pumpkins and pulled up their vines. Jānis dug parsnips, and he and Dainis pulled up carrots and cleared their own small gardens. I harvested broccoli, fennel, dill, and corn, and then washed and stored the root vegetables, cleaned the broccoli and fennel, shucked corn, and piled squashes on the porch for curing. Let the gales of autumn come. Our harvest is in.

Now rain lashes the house windows, and wind whips birch and apple trees in the front yard. I'm beat, but once again a gardener's satisfaction of outrunning the storm lifts and carries me on into the wuthering night.

October 5 (-3°C, mostly clear, calm, 7:45 a.m.)

A vireo sings a few notes from shrubs beside Balmoral Road. Juncos skim over the pavement, voicing "tsuk-tsuk-tsuk" in quick conversation at my passing. I'm looking for deer, for horizontal lines among vertical trees, for perked ears and flashing tails. Three mornings ago, I saw a buck. I told myself to look across the pavement to see what I could see, and there he was, strolling across the access road to an abandoned homestead and the clear-cut behind it. The buck raised his antlered head in inquiry. He and I gazed at each other for perhaps half a minute and then peaceably continued our separate ways. All spring, summer, and autumn up until that morning I saw only does and fawns. When I mentioned the buck to Vilis, he told me of how, when he sat in a tree blind in southwestern Manitoba with a bow in his hands, he watched does and fawns walk past him in plain view whereas bucks moved cautiously along "sneakier" trails (his description).

A seed-filled scat lies humped at the edge of Matheson Brook Road. Thick, purplish, and pimpled with pale seeds, it looks very much like the raccoon scats in our coop. Vilis trapped the coon four days ago, and he and the boys drove fifteen kilometres before releasing it.

A ruffed grouse blasts across the gravel road. I see the blur of its wings but don't hear their whir. Sunshine warms my back, and then I'm in shadow and the air is cooler, sweeter, the roadside grasses frosted in deep pockets of shade or beaded with dew where ribbons of sunshine play over the road. Asters still bloom white and mauve at the gravel's edge. Senescing bracken ferns are shades of cinnamon and gold.

A few steps farther on, our front road is a beckoning tunnel of green and yellow, with wild apples underfoot and the brook sliding smoothly beside me. I know now, as I never knew before this autumn, that vertebrate life exists beneath nearly every stone on this slope and under many pieces of fallen bark. The world of salamanders has opened to me.

Yesterday, my sons, two young friends, and I came to this steep damp slope above the brook to search for salamanders. We carefully lifted large stones and fallen branches, with our gazes fixed on the soil surface revealed beneath them. Our wits sharpened as we developed search images for slender shiny bodies with moist skin, side-splayed legs, and long tails. Excitement raced through us when we found the first of several trim red-backed salamanders, each with an orange-red stripe along its back. We exclaimed over yellow-spotted salamanders that looked as though a child had daubed dots of bright yellow paint on their blue-black skin. We even found a minute blue-spotted salamander, its glossy black sides speckled with dusty blue. The tiny blue-spot's legs were so thin and delicate I feared they would break as it clambered over my hand.

I'm jolted back to the present by the sound of water dripping onto metal. This is an unexpected melody, one produced by robins knocking moisture from pincherry leaves onto our overturned canoe.

Two days ago, Dainis, Jānis, and I dragged the seventeen-foot aluminium Grumman down the hill and launched it on our pond in search of amphibians, the subject of our latest intensive study (hence the salamander search yesterday). We juggled paddles, dip nets, and trays and took turns paddling and dipping. Several times we ran the canoe into shrubs at the small pond's edge before we learned to change direction more quickly. With experience, we also learned to glide quietly with paddles poised, ready to brake and thrust the canoe backward or to shoot it ahead. We developed search images for bulging frog eyes above the water surface and for motionless frog bodies at the pond's edge. We learned to creep up on those frogs on shore and to guess at the trajectory their leaps would take when we went for them with the dip nets.

We netted three green frogs, each a different hue—one, brilliant lime green; one, olive; one, almost turquoise. We even scooped up a red-spotted newt from the muddy pond bottom, its olive back dotted with scarlet circles. The frogs tried to escape from the trays, and the newt swam wildly. We quickly keyed them out and returned them to the pond. One of the frogs hung in the water, watching us for what seemed like minutes before it kicked and dove away, the pond water deepening and darkening over its extended legs. We paddled to shore, tied the canoe's bow line to a tree and carried dip nets, trays, and our new-found knowledge up the hill.

October 8 (-3°C, clear, NW wind, 7:45 a.m.)

It's good to walk the back road again, to rest my spirit in hemlock woods. For the past two days I was away, teaching creative writing workshops to elementary school children on Cape Sable Island, that small hourglass of land off Nova Scotia's southern tip. The landscape on that coastal island—sand dunes, salt marshes, beaches, sparse stands of conifers—was so different from these woods and our meadows. The human landscape was different, too—fishing boats, lobster pounds, fish-packing plant, sunrise fishing conversations at the Tim Hortons on the other side of the causeway.

After my first day of workshops, I walked a beach with binoculars around my neck and bird guide in hand. Semipalmated sandpipers rushed into the surf, their black beaks bobbing and their black legs blurring. A greater black-backed gull flew over the ocean, its wings, angled contrast to the rolling curves of incoming waves. Three female common eiders—big cinnamon-brown sea ducks with sloping bills—rode waves beyond the surf. At my feet, lumpy kelp corpses lay tossed onto the sand by the Atlantic Ocean, their twisted stipes resembling neoprene tubes, their shiny blades, broad and frilled like lasagne noodles. Among the kelps, a stranded alga with long fine fronds lay spread across the sand like green hair streaming toward the water.

I met a fisherman's son. He told me of seals he'd seen off the rocks farther up the beach, of how fishermen sometimes shot seals, and that it was ducking season. I walked on toward the rocks and collected ocean sponges cast ashore for Dainis and Jānis, the coastal wind cold in my face. I reached the rocks and continued past without spotting any seals.

As I turned back, disappointed, I caught a glimpse of something sliding off a rock. And there they were—one, five, six, ten, a dozen, maybe more. One seal yawned and leaned into a back float before slipping beneath the waves. Others appeared to be treading water, their snouts, dark and silver in sun and shadow, pointed to the sky. I spotted a thick body on a rock. Another swam to tease or challenge it, and then both were gone. All were gone.

I tucked the sponges into an empty crab shell, feeling the stickiness of ocean salt on my fingers. The Northumberland Strait is only a fifteen-minute drive from Ravenhill and every year Vilis, the boys, and I walk its sandy shore, splash in its waves, and swim in its summer-warm water, yet always, *always*, its ocean salt, like the salt on my fingers when I collected those sponges, comes as a surprise. We are where we came from, and I hail from the prairies where alkali, not ocean salt, lined the shores of Thomas Lake. Had I grown up with the ocean on my doorstep, the feel of its salt on my fingers would be as familiar to me as that of alkali's dusty white powder.

While I was away, Dainis and Jānis had more amphibian adventures, telling and writing stories about frogs, newts, and salamanders. They also had a startling bird adventure. A ruffed grouse crashed through an east window of our house, killing itself and sending splinters of glass exploding into two rooms less than ten seconds after Jānis finished sharpening his pencil and ran out of one of those rooms. I thank God for mercies, for Jānis's quick feet, and that I'm home.

October 11 (3°C, mostly cloudy, strong NW wind, 9 a.m.)

Gold lies on the back road, dropped by yellow and white birches. More gold hangs from twigs alongside Balmoral Road and is strewn in ditches. Beside Matheson Brook Road, red maple leaves flame against a backdrop of evergreens, and ash seeds dance like brown fringes of a dress. Clarion calls ring out as Canada geese wing overhead, their dark wings whipping air, their white cheek patches bright in the morning sunlight. There are times when geese honking in flight, and golden leaves fallen like coins to the ground, are riches beyond measure—when just being is enough.

October 12 (Cool, overcast, 3 p.m.)

Today my family is hiking in dappled sunlight beneath stands of sugar maples in Wentworth Valley, a half-hour's drive west of Ravenhill. Roots and stones press into my feet on hilly trails that stretch my calves and tighten my knees. Jānis romps ahead on the trail, his tufty hair bouncing, his feet kicking up leaves that he scoops into his hands and throws at his father and brother. Vilis and Dainis retaliate by stuffing leaves down Jāni's shirt, and then we all throw leaves at each other, stealing handfuls of autumn and tossing them in bright echoes of their earlier falling.

This weekend is Thanksgiving weekend, a time to celebrate the gifts in our lives and the bounty the earth provides. Our feast tomorrow will be a little different, since it will be the first since we moved to Ravenhill that doesn't include cranberry sauce made from berries that we, ourselves, picked. Every other October, whether in wind and cold or sun and warmth, we made the long drive to Polly Cove on the rugged boulder-strewn landscape of the West Dover–Peggy's Cove Barrens. There, on the glacier-scraped spine of Nova Scotia's South Shore, we scrambled over rounded granite outcrops and collected wave-strewn or gull-dropped treasures of sand dollars and sea urchin tests. We marvelled at pitcher plants and, of course, searched for plump cranberries hanging amid miniature *Vaccinium* leaves and sphagnum moss. All around us, glacial erratics—gigantic boulders once carried by a glacier that melted 12,000 years ago—loomed like frozen sentinels forever on display against sky and sea.

The barrens are a wild landscape, a hard landscape, a landscape of rock and crashing waves and hardy plants clinging. We go there as much for the wildness as for the cranberries, and when we're home again and eating those cooked and sweetened cranberries, we taste the barrens—the rock, sea, and wildness. Wentworth Valley has a different wildness. It's softer and more elegant, with slim trunks, tall trees, and orange flaming on its breast.

The barrens, this maple forest at Wentworth, our own property—all allow my family to experience some of the earth's tremendous diversity of landforms and plant and animal life. Such diversity feeds the human spirit. Each new landscape we experience turns a new page in our lives and opens our minds and senses to the myriad marvels of Creation.

When I flew north to Alexandra Fiord for the first time in late June of 1980, it was like leaving my past behind and landing in a new world. Gone was the lush green of Nova Scotia in early summer. I jetted northward over progressively more barren landscapes, then, in a chartered Twin Otter, arrived in a land of rock and ice. The Otter made a rough landing on a sand airstrip squeezed between two lowland rivers. Heart pounding, I handed my terrified dog to another researcher on the strip and then scrambled out of the plane. Frigid air swamped me as I watched Skyree run to a snowbank, roll in it, and then begin sniffing at tundra plants.

Initially, I was out of my comfort zone, startled at the immediacy of the rugged landscape—the treeless lowland, the frozen fiord, the mountains hung with glaciers. Yet, an excitement unlike any I had experienced before raced through me. There I stood on a High Arctic lowland in one of Canada's most remote locations, nearly at the top of the world. I'd be lying if I said that this realization didn't tinge my excitement with uncertainty, even with fear. At the age of twenty-four, I had embarked on the greatest and potentially most dangerous adventure of my life. Could I handle the isolation? Did I have the skills, insight, and perseverance to complete my research? The answer to both questions was yes.

A new place, a new environment, always tests us. It challenges our knowledge of who we are and might awaken aspects of ourselves that we didn't know existed. Alexandra Fiord taught me to believe in myself. The Polly Cove barrens taught my sons to reach for wild places. And this elegant maple forest in Wentworth Valley teaches my family that nature's beauty is ours to relish if we but reach out a hand to catch its falling leaf.

October 13 (morning)

I'm walking the back road, serene, when sudden fear catches my breath. A dark shape rises from the forest floor. But it's only a stump, and my fear subsides. I'm surprised. I didn't expect this reaction. Even after eighteen years, I still see bears in stumps.

During the summer of 1979, before I left Alberta to pursue graduate studies in Nova Scotia, I had a job trapping small mammals in the Northwest Territories. The closest real town was Hay River, but I was based at Heart Lake Biological Station, a cluster of log buildings perched atop a limestone escarpment that overlooked a vast expanse of boreal forest to the north. One of my two main tasks was to set snap-trap lines in various habitats to sample populations of small mammals, mostly deer mice and red-backed voles. My other major assignment was to set and check two live-trapping grids, the aim of which was to tag small mammals in two specific populations, one living in an upland mixed forest, and the other in a mossy spruce forest.

During the first week after my boss, Bill Fuller, and I opened the mixed-forest live-trapping grid near Heart Lake, a black bear just out of hibernation found the shiny aluminium Longworth traps we'd set and pawed and chewed them open to get at the mice and sunflower seeds inside. "Doc" shot that bear and grumbled that it had been a long time since it had taken him two bullets to shoot anything. A tall, lanky graduate student and I skinned out the bear and butchered it while my puppy, Skyree, growled at the smell of bear. Such was my first experience with bears.

My boss headed back to Edmonton, leaving me in charge of the trapping. I thought I was bear-free, until the next time I opened the same live-trapping grid and found a new set of bear tracks walking up and down the grid rows. Again, the expensive two-piece Longworth traps had been broken open and punctured with tooth holes. I even spotted the bear plodding peacefully along a grid row under tall birches to the west of where I stood assessing the damage. The young man with whom I'd skinned and butchered the first bear bought a licence and shot the second bear on the grid one calm evening while I waited nearby, blood running down my lip from the unconscious bite of my teeth.

Again, I thought I was bear-free, but bear whiskers were caught in a snap-trap, skunky bear smell hung in the air, and I saw a bear beside the road, a bear high in a tree, a bear rearing up by the shore of Heart Lake, a bear prowling along my snap-trap line while I checked it—snowshoe hares flushing ahead of the bear, Skyree whining and running to me. And then, there was a bear sow with three small cubs who found the live-trapping grid at Heart Lake. One evening while I checked the grid with another assistant, Kevin, the sow and cubs raced across the path ahead of us. The mother bear sent the cubs up a tree but remained on the ground. We could hear her moving in a dense thicket of young spruces but couldn't see her. We backed away, sweating, not knowing exactly where she was or if she would attack.

The graduate student arranged to borrow a huge metal live-trap to capture the bear family, and we hauled the trap by truck to near the live-trapping grid and baited and set it. When we checked the trap, the door was closed, so we thought we'd succeeded in capturing the four bears. Then we heard crying from inside the trap. When the student walked toward it, the sow exploded out of nearby shrubs, causing him to race back to the safety of the truck. It was the worst possible scenario—hot sun beating down on the metal trap, hungry cubs inside, distraught sow outside. With the sow poised to defend her cubs, it was impossible to open the trap safely to free them.

The student telephoned the local Conservation Officer for advice, to see if a tranquilizer dart gun was available to immobilize the sow, or if there was a zoo or game farm that would take the trapped cubs. The answer was no, so he did the only thing he could. He shot the bears—first the sow, to protect us, and then the cubs. He later told me that the cubs were crying like babies when he opened the trap door.

For the remainder of that summer, I saw bears everywhere—in every stump and behind every pile of deadfall. At times, I was so afraid, I shook. At times, I pleaded with the graduate student to accompany me while I checked the grids. I don't know how I would have made it through those months without him. At the end of the summer, he and I went our separate ways. He returned to Edmonton to write his thesis, and I drove to Halifax to begin my own research. I see him often now, though. He's an ecology professor at the Nova Scotia Agricultural College, and the father of my children.

I walk on, leaving behind the dark stump that jolted a past fear into my present life. After that summer at Heart Lake Biological Station, I thought I would never again choose to work in bear country, yet the very next summer I strode over High Arctic tundra and watched a polar bear hunt ringed seals on Alexandra Fiord. The shadow of that fear still loomed over me at "Alex," but my time on the lowland offered rewards far more profound than that shadow.

We can only move forward from fear, and in the going, grow in experience, strength, and wisdom. Here in the backroad woods, I step into the new day, seeking rewards that lie ahead and are more profound than the shadow behind.

October 22 (-6°C, mostly clear, calm, 7:45 a.m.)

My breath is white heat against cold air, the meadow grasses immobile under frost. In the woods, fallen leaves are crisp flakes of autumn beneath my feet. I seek them out, make them crackle, and then return to the middle of the road where footsteps fall more quietly. The Cathedral hemlocks rise tall around me, straightening my spine and pushing the crown of my head up into the sky. They are as still as a breath held, as dark as midnight. Beyond them, in the thinned forest near Balmoral Road, leaves of maple, birch, ash, and aspen hang like autumn flames frozen in still air.

My clothes and footsteps are too loud for this morning. On Balmoral Road, an oncoming car is a locomotive bearing down on me and then receding. An ATV's roar is an explosion against the hillside. We've made our world so loud. What would I have heard had I walked here a hundred years ago? The rumble of nearby Balmoral Grist Mill and the jingle of horse bells? What about two hundred years ago? Bird song and light steps?

On Matheson Brook Road, autumn is aflame, its brilliant tongues, the shivering leaves of aspens, bright in the morning sun. A ruffed grouse blasts across the road, the dark band on its broad tail, one arcing curve, and the curl of its braking wing, another. A nuthatch voices its nasal "meep," and leaves whose petioles clung to twigs throughout the night give up their hold and fall lightly to the forest floor. The neighbour's dog barks. A school bus passes. The loud sounds are here, then gone, and again I hear the soft fall of dying leaves. Chickadees murmur.

My breath is white heat against cold air, the roadside alders immobile under frost.

On our front road, fallen leaves crunch underfoot. Sunlight that doesn't yet find this dark, shaded slope hits sugar maples and largetooth aspens on the far side of Matheson Brook. They flame with golden fire while I walk in frost.

The red maple by the pond stands naked, her dress in dull tatters on the ground, her grey arms reaching to the sky. Old grasses clutter the hill. On the porch, pumpkins and dry cornstalks are shades of orange and gold. In the house, I smell apples drying, eat stone-ground oatmeal, and paint my mind with images of orange and gold, of grouse and flames, of dying leaves and the white heat of breath against cold air.

October 25 (-3°C, mostly clear, W wind, 10 a.m.)

At the edge of the backroad woods, out of the sun, cold air is like chilled hands on my face. Here, the road's frosty surface is spotted with imprints of aspen leaves that fell to earth, rested, and then were scattered by the wind. I see no frost ahead in the forest. Always, the meadows freeze first, and always the forest freezes last. And this pocket of land at the forest's edge, which is neither meadow nor forest, is like the portal to another world. I step into it, and new sensations engulf me as in no other place on this road or on our land. In this place of change at the edge of the woods, I'm in transition, caught between two habitats yet immersed in another that has its own distinct character.

Always the meadows freeze first, but they always thaw first, too, their rolling chests bared to the sun. Always the forest freezes last, but it always thaws last, too, the blanket of conifer trunks and branches stacked upon its chest as loath to let in warmth as to let in cold. The thought occurs to me that perhaps people are like that, too—like meadows and forests. Some of us, like the meadows, bare our hearts wide open and wear sleeves stained with blood from those torn hearts. Some of us, like the forest, hide our hearts beneath trunks and branches, as reluctant to let in warmth as to let in cold. Where is the transition? Where is the place of change at the edge of the woods, the place of change between not love and love? What is the portal that leads from one into the other yet is a separate state?

Fire flares at the top of the back road, its flames every point of maple leaves that dance in a pile on the gravel. The leaves' red, orange, scarlet, and coral make me want to write, not essays or articles but rather letters to family and friends, as though this brilliant shedding of colour were a signal to reach out before snow and ice arrive.

A doe crosses Balmoral Road ahead of me, and I think she may not even be aware of me as she steps onto the gravel access road where I spotted the buck. She would suit that place of change at the edge of the woods, for deer are creatures of the edge—edge of forest and field, edge of night and day.

On Matheson Brook Road, two boys zoom past me on an ATV, followed by another on a dirt bike. All three hunch over their machines, fiercely intent on speed.

Speed is a heady thing. When I was a teenager, I rode my brother Tod's horse, Tibbio, a stubborn and undisciplined sorrel mare with thick muscles and a wildly exhilarating gallop. Tibbio and I made a pact. For her part of the pact, she carried me a kilometre or more away from the farmyard, over the grey dusty track through grain fields—where my family sometimes walked with our cats—and then along the shore of Thomas Lake. For my part of the pact, I let her run home.

I recall the creak of leather as we left the farmyard, and my gentle rolling in the saddle, as well as the thrill of anticipation. At the far edge of the farm, Tibbio began to prance, knowing her time to run was near. When I neck-reined her into the turn, her muscles bunched and I clung to the pommel as she nearly burst out from under me. Wind stung my face. Wheat fields and hay fields blurred past. The smell of hot lather filled my nostrils as Tibbio galloped like the wind, and I—caught up in the wild immortality of youth—rode her for all I was worth.

Tibbio is long dead, and I now live more than half a continent away from the farm of my youth. I don't own a horse and haven't ridden in years. Still, when the barrel racers enter the MacMillan Show Centre in Bible Hill during the Nova Scotia Provincial Exhibition—their mounts prancing as they await the signal to turn and race—part of me rides Tibbio again. Our muscles tense as she bolts into the turn, and then there's nothing but wind sting and power, golden fields blurring past, and the smell of horse lather as I hunch over her withers, fiercely intent on speed.

Up the hill, sunshine warms our south meadows and melts frost brought by an overnight low of -8°C. Snow fell on Ravenhill yesterday, a light skiff that melted in morning sun and lit an anxious fire in me to complete outdoor autumn tasks. In Nova Scotia, there's no telling when winter will arrive—perhaps in mid-November or perhaps in January. It might fill the yard with deep snow or spread it with a sheet of ice. It might send temperatures plunging to -30°C or keep them bobbing around the freezing mark.

Vilis and the boys join me, and we spend the afternoon working outdoors. At dusk, a section of lawn is newly mowed, two small flowerbeds lie brown and freshly spaded, and our porch is lumpy with bags of baking apples, drying apples, and dahlia tubers. On this Saturday, with my family around me, the weight of autumn work is a leaf lifted by the wind.

October 27 (-6°C, overcast, calm, 7:40 a.m.)

This is a fresh morning, a good morning, a morning in waiting for the sky to split open its belly and spill its guts of rain or snow down onto me. Frost's hoary fingers grasp grasses and goldenrod on the front-road hill. Spears and plates of ice fleck the pond's surface. Cold slides into my jacket and creeps over my shoulders. I curl my fingers and thumbs into fists in my mittens to ward off chilled air.

The old bridge is slick, as I knew it would be, and I marvel at the brook's low water level. We've had no rain to speak of for a month—not since the day I tilled the garden. Potato farmers on Prince Edward Island, who last autumn were unable to harvest their potatoes because the soil was too wet, this fall are unable to harvest because the soil is too dry. Dairy farmers are also suffering. A neighbour whose fields produced far less hay than usual has only enough to feed his dairy herd for half the winter and already sold his beef herd.

I place outgoing mail in our zebra mailbox and cock a black plastic ear to catch the mail driver's attention. Then I walk the loop in reverse, gathering litter as I go. Discarded bottles, beer cans, plastic bags, and Styrofoam cups soon fill my trash bag. More litter remains, but I've done what I can today, and even this small amount makes a difference here on Matheson Brook Road. Maybe it'll make a difference elsewhere, too. Mathematicians debate whether our universe is ruled by equilibrium or chaos. Do random disturbances eventually cancel each other, and whole systems return to equilibrium in a predictable manner? Or do single random disturbances, insignificant in themselves, change future sequences of events in ways that can't be predicted and that are, therefore, chaotic? The chaos theory has been likened to a butterfly flapping its wings and influencing events on the far side of the world. I like to think that my collecting this one bag of litter—an act not necessarily significant in its own right—will have ramifications that spread throughout the world.

It's odd to walk the reverse of my usual route. When I reach the back road, I feel as if I'm going in the "out door," so to speak. A dozen steps down the road, I pause to drink in the sight of a red maple standing tall beside me, its crimson leaves motionless and achingly lovely against mottled grey bark. When those leaves fall, a year will pass before I see their vividness again.

Seasons, those times of renewal and change in the natural world, also flow through our years like water. We never really know what lies ahead in the spring at the end of our winter. I leave the woods—green and sheltered—and enter our meadows—white and frosted. This is a fresh morning, a *good* morning. If this is autumn, let it not end too soon.

October 28 (7°C, overcast, strong SW wind, 6:45 a.m.)

Puddles in the backroad woods hold trembling images of trees and sky. I lean over and gaze into one until my reflected head and shoulders appear. As a child, I became dizzy when staring into puddles. I felt as though I was falling into the sky and pulled myself back from the edge with a pounding heart. As an adult, I remember the sensation, but now the feeling isn't strong.

Last night's storm, during which the roiling sky poured rain onto us, took many leaves. Now they blanket ditches and dance in rustling scrapes across Balmoral Road. All around me, as I remove my jacket, hat, and mittens, the wind's song is sung forte, with crescendos and decrescendos that roar and whisper among trees. If yesterday morning was one of waiting, this morning is one of motion and change. Tamarack needles have started to turn their conspicuous and surprising yellow. Matheson Brook runs stronger and speaks louder, its pool beneath the bridge, deep, dark, and wide. Burgundy features prominently in blackberry leaves, wild strawberry leaves, and small sour apples that hang like Christmas ornaments on the old tree in our front yard under whose boughs we buried Skyree. A snow fort that Dainis and Jānis built after the snowfall four days ago rears up in an incongruent lumpy circle on the lawn, its blocks of half-melted snow flecked with green grass clippings. Winter solstice won't arrive for nearly two more months, yet this morning's dynamism holds the sense that we're perched on the edge of winter.

Change is always dynamic. A new start is inherently action, even if only in the mind. We are all authors of our own life stories and, as such, have a big say in shaping our story's protagonist and deciding which events will comprise the plot. In the world of fiction writing, some authors are plotters; others are pantsers. Plotters craft detailed outlines for every chapter in the as-yet-unwritten novel. Pantsers know the beginning and the end, probably a few scenes in between, and fly by the seat of their pants, working out the story elements as they go.

In writing and in life, I'm a pantser. The book you hold in your hands opened with my protagonist walking an old wood road beside Matheson Brook, seeking who she was. She began writing a journal to record her observations and reflections and, in doing so, cast herself on the waters of discovery, both external and internal. She swam and waded and haltingly dipped her toes in those waters, found pearls of immeasurable worth, and sought ways that she might shift and change and see things she'd never seen before.

The story—my story—is probably about half done. I have no idea what will happen in the rest of it. I'll work things out as I go. The future will come, and the future always brings change. In a piece of fiction, the protagonist must show signs of growth and change in response to the story's events. At the end of this journal, how will I have changed and grown? I have no outline to follow, only the intuitive awareness that these months of observation and reflection will add new dimensions to my character, bring clarity to my vision, and offer—like a pearl of immeasurable worth—peace.

October 30 (-3°C, overcast, N wind, 6:40 a.m.)

In the east, hills are grey curves pushing against a pink-streaked sky. Today I walk with the joy of one about to be tethered. Rifle season for deer opens tomorrow. Yesterday Vilis sighted in his shot gun and bought slugs and a licence. He radiates excitement. So do Dainis and Jānis. They want to eat venison and make deer-hide drums.

I'm not excited. I see only five long weeks stretching ahead, locking me into my house and yard except on Sundays, when hunting is not allowed. I could go out—I own a hunter-orange vest and toque—but during rifle season I can't make myself walk or run through the woods at dawn before Vilis leaves for work. Here's why: I was working as a resource aide in Boissevain School in southwestern Manitoba on the day four children were told that their father had been killed by a hunter—shot straight through the chest while checking his trap line one frigid November morning. So, I don't walk or run in the woods at dawn during rifle season and I take nothing—especially the gift of my life—for granted.

November 9 (7°C, overcast with drizzle, calm, 9:30 a.m.)

An earthworm stretches across an expanse of wet gravel at the yard's edge. Beyond the yard, rain-beaded heads of meadow grasses droop white beneath their loads of moisture. The scene is soft and fuzzy, a plush bedspread draped over the land's curves and interrupted here and there by sharp upward thrusts of goldenrod. In the backroad woods, water drips from trees, and I skirt wide puddles. A drop splashes onto my forehead—

a christening, a baptism. Today is Sunday, and the woods and roads are mine again.

Fallen leaves at the top of the back road are piled in brown windrows, their fire faded, their substance leaching and crumbling into food for the decomposers that feast on them and set the stage for next year's flush of new green and autumn fire. Balmoral Road's verges are alive with sparrow and chickadee voices, and a hairy woodpecker swoops across my field of view before landing low among an apple tree's gnarled branches. Hallelujah! This world is still here, still full of life, sound, and texture.

Water from recent rain lies in brown strips in the ditches alongside Matheson Brook Road. Old man's beard lichen hangs like pale green tinsel from evergreen boughs. A nuthatch's "meep" sounds close by, and a raven's gruff voice reaches southward from the north, telling me that even here in northern Nova Scotia, where houses line much of the road between Truro and Tatamagouche, wildness exists.

And the raven would know. It was a raven that stole my groceries from the back of a pickup truck in Hay River on the shore of Great Slave Lake. It was a raven that picked clean the bones of a wolf-killed moose beside the Alaska Highway near Kluane Lake in the Yukon. It was a raven that spread its wings—big and black—above me while I walked the High Arctic tundra under the midnight sun at Alexandra Fiord on Ellesmere Island. The raven knows wildness, it knows wilderness, and when I hear its hoarse voice calling, I'm in those places again and I know wilderness, too.

Hours later, I walk the back road with my men before dusk. Damp beech leaves alongside the road are beaten copper on wrought-iron twigs. Mist hangs, white and still, among slate columns of hemlock. We might be somewhere on Canada's West Coast, except that we don't have salal to thrash through in these woods. The damp, misty air is the same, though, and I'm half here, half there, walking beneath Douglas firs and western red cedars near Victoria, or among arbutus trees with bare limbs raised, crooked and coppery, their bark hanging like grey rags in the mist.

Dainis and Jānis kick up packed wet leaves at the roadside, and Jānis says how much fun it would be to ride his bike through *that* big puddle. Blotchy joins us, her ears flicking nervously at the sound of cars on the pavement ahead. Wisely, she turns away, leading us back along our quiet access road.

We leave the misty woods and marvel at the variety of browns in our rain-washed meadows—chocolate and cinnamon, taupe and tan, rust and a shade only a little browner than cream. All are intensely rich in the way I'd forgotten browns can *be* intensely rich. I see no pinks or purples, no blues, brilliant yellows, or emerald greens—hues that belonged to spring and summer—yet the meadows are vibrant with colour, filled with shades of the earth. I want to lie among them and wrap them around me, wear them in my hair and dangle them from my ears.

The earth isn't the only thing that reveals riches in unexpected ways. Last evening, my sons surprised me. It was dark, after supper, and the kitchen was filled with pumpkins and apples and the voice of Leonard Cohen singing "First We Take Manhattan," from his *I'm Your Man* album.

Dainis looked up from peeling and chopping pumpkin. "He has a fierce voice."

Jānis tossed pumpkin peels into the compost bucket. "He sounds like Darth Vader."

"Yeah. It's a fierce song." Dainis continued chopping. "I like it. What's it about?"

I placed apple rings in a food dehydrator, talked of poetry and images, and Vilis explained how music doesn't always have to be understood to be appreciated. Inside, I was laughing, thrilled and a little shocked that my sons liked "fierce" music. Next up was "Ain't No Cure for Love."

Again, Dainis looked up from his task. "I've heard that before on the radio, once or twice."

Just for fun, Vilis and I sang the chorus along with Leonard, belting out the lines about doctors finding no cure for love.

"Why would they want one?" That was Dainis again, with all the wisdom of a nine-year-old sage whose heart is torn all too often from wearing it right out there on his sleeve.

When "I'm Your Man" came on, I told Vilis to stop peeling/coring/slicing apples and dance with me. He washed his hands and followed me into the living room. At first, we forgot how to do it—that he should lead, and I should follow. We stumbled and went in different directions, but then our feet and bodies remembered, and we moved together, dodging the couch, the table, the rocking chair. We even did a twirl before Leonard's last growl faded. Then we thanked each other and returned to the kitchen, to the pumpkin and apples, to the watching eyes and knowing smiles.

Fierce voice. Young sage. Why would anyone want a cure for love?

November 16 (-2°C, overcast, calm, 8:45 a.m.)

New snow crunches underfoot. Snow-laden grasses bow to the ground. This morning Ravenhill is a pen-and-ink drawing of spruce trees on white meadows, with the back road's gentle curves winding through the bucolic scene. At the place of change at the edge of the woods, curved tracks show where an incoming vehicle turned around yesterday—likely a hunter.

Silence reigns in the forest. Snow-crusted hemlock boughs hang like rag dolls. I catch a flicker of movement—a red squirrel—but even the squirrel is silent, as though sound has no place in this stillness. Mother Teresa wrote: "In silence we will find new energy and true unity. Silence gives us a new outlook on everything…In nature we find silence—the trees, flowers, and grass grow in silence. The stars, the moon, and the sun move in silence."[40] In my harried existence, silence is a blessing. Rather than evoking emptiness, it evokes peace and the anticipation of what lies beyond this moment.

Earlier this morning I saw, from my bedroom window, a white-tailed doe and her two fawns eating fallen apples under a tree in the south meadow. Their heads lifted and dropped, their legs shifted, and the meadow lay around them like a Christmas-card scene. What did the fawns think of snow and cold, of fallen leaves and frozen apples? Yesterday morning, before breakfast, my own two "fawns" raced outdoors and rolled sticky snow into giant balls before positioning the spheres in a circle as the foundation for a new snow fort. Their gloves, coats, and pants got wet, but Dainis and Jānis didn't mind. Giant smiles lit their faces, their cheeks rosy from fresh, cold air. Did the white-tailed fawns also play yesterday? Did they sniff the snow and flick their ears? Did they prance and leap?

On Balmoral Road, I hear hushed sounds—subdued sparrow voices, the faint call of what might be a crossbill, and distant squirrel chatter. Two small birds, perhaps crossbills or redpolls, race across the sky, diving, lifting, weaving in perfect formation until they're lost against a backdrop of spruce trees. Nailed to the dazzling white trunk of a paper birch beside the pavement, a sign has bullet holes in it. Three weeks of rifle season remain.

Beside Matheson Brook Road, pine grosbeaks—the males, pinkish red; the females, mustard yellow—tug at ash seeds hanging brown against a white sky. Rose hips—red and lush—startle me with their unexpected colour and roundness in an otherwise slim-line landscape of grey stems and trunks lined with white. I notice deer tracks in the snow on the road's verge, some fresh, some likely made yesterday. We heard gunshots from this direction late yesterday afternoon, three in succession. As I look up from the tracks, a doe bounds across the gravel—a horizontal brown slimness between vertical white spears of roadside shrubs.

The forest's silence surrounds me, as does the scene of bare uplifted branches edged in silver, and of thickly needled boughs laced with white. Peace pervades this Sunday morning, peace and a sense of release as though permission has been given to let go of things undone and begin anew in a new season—winter.

Our lives are filled with new beginnings. All are steppingstones that lead us onward into our future. At times, a turbulent river may roil around the stones and catch at our feet, threatening to swamp us. At other times, the river runs smoothly, peacefully, and all is golden, the stones firm beneath our feet. Each new stone takes us further from our past, yet the past doesn't let go easily. Sometimes, as Lucy once said in a *Peanuts* comic strip, the past gets in our eyes. It blinds us to where we are in the present, to *who* we are in the present, and to the countless phenomena that surround us. The Loggins and Messina song "Watching the River Run" talks of moving further away from past actions and leaving each one behind, yet only beginning to watch the river as we listen, learn, and yearn. Often when I walk our front road beside Matheson Brook, that song plays through my mind.

Now, a chickadee calls quietly from the forest alongside Matheson Brook Road. On our front road, I spook a ruffed grouse that sputters and fusses as it walks away, its jerk-jerk footsteps rustling among brown leaves. I cough, and the grouse flushes from the ground, arrowing straight across the brook that speaks louder now, stronger with yesterday's snow and rain. A beaver has been at work near our bridge, pushing branches into the stream bed. Already the waterway is wider and deeper, its surface swirling only an inch below the upstream edge of the bridge's sagging deck. We'll see what winter brings, with its storms that whip the land and pour water down hills until Matheson Brook is a brown fury that floods our pond, spits over the bridge, and punches holes in beaver dams.

My shoes are drenched, and my socks and pants whitened from snow falling from every leaning goldenrod and drooping grass stem that blocks my path up the hill. In our yard, small apples still hanging on a front yard tree are rich red candles against the grey and black of bark and the white of snow. I picture a blue jay among the apples and determine to see one there. To that end, I will issue an invitation of black sunflower seeds and wait.

November 21 (-3°C, clear, light breeze, 9:15 a.m.)

Sunlit branches of honeysuckle, willow, aspen, and pincherry are shining streaks on somber grey, like colours revealed when as a child I scraped away black ink painted over a crayoned background to reveal hidden streaks of blue, green, pink, and orange.

In the yard chickadees twitter, and evening grosbeaks—the males resplendent in their black and yellow plumage—voice single note calls. A blue jay claims the bounty in the pole feeder, and our speckled rooster scratches and pecks up seeds under the feeder before striding quickly to the coop. I opened the coop this morning to allow our flock to enjoy the sun and warming temperatures and to scratch at fallen bird seed and frozen vegetable wastes tossed atop the compost heap. It's mid-afternoon now, and Samson and four hens preen near the spruce windbreak, basking in sheltered warmth reflected from our old shed's metal doors. One hen settles into sun-dried grass and stretches her leg out in a moment of abandoned luxury.

All our original six Golden Comet layers have passed on. The oldest, Redwing, died at three-and-a-half years of age one September morning last year. When I entered the coop, I found her standing still and hunched beside the feeder. As I watched, she squawked and fell over, her wings flailing, her feet scratching at nothing, her neck stretching out, her eyes closing. I left the coop, walked as far as the woodpile, and then turned back. Redwing's legs were still, and her neck, curved backward. One wing reached outward in a last sustained curve before settling in slow motion against her body.

So, chickens come, and chickens go, yet we reserve a place in our memories for the first ones and the special ones—for Goldie, who was so tame the boys could carry her around like a kitten; for Colour-Cluck, who was broody and feisty and the best mother hen we ever had; and for Samson, who walks as if he owns the earth.

Our ruling rooster abruptly leaves his basking hens and races across the yard to the coop, no doubt with the intent of chasing his son, the speckled rooster, away from our other two layers. The speckled rooster is the underdog in these confrontations, but he's fast and sneaky, making dashes for unattended hens, mounting them swiftly and roughly, then hopping off before his father arrives to knock him off.

The year we built the coop, when our Meat King chicks leapt with joy on being released from the stock-watering tank, we also raised turkey chicks—three of them. To remind ourselves that the turkeys were food animals, not pets, we named them Thanksgiving, Christmas, and Just-In-Case. Just-In-Case (whichever turkey kicked the bucket first) died at a week of age, but Thanksgiving and Christmas grew into stately white gobblers that, come autumn, flared their tails and danced at each other in challenge. When spooked, they stretched their long pink legs into ground-eating strides and lifted briefly into the air like heavily laden Twin Otters.

I became nervous when the gobblers began to show their aggressiveness and when they grew to a height at which their beak level corresponded with little Jāni's eye level. I suggested to Vilis that we butcher the free-ranging birds soon and instructed the boys to scare them away when they came near. "Rah!" Dainis would yell, his hands darting out as if to push the turkeys away from his squealing brother. "Rah, rah!" And off the Twin Otters would fly.

We butchered the turkeys a week before Thanksgiving Day. Thanksgiving dressed out at twenty-one pounds, and Christmas, at twenty-four pounds. As we sat down to our Thanksgiving dinner, Vilis carved the roasted turkey and served each of us a portion. Dainis stared down at his plate, not picking up his fork. Vilis and I watched him, wondering if he would say no, he couldn't eat *this* turkey. Then a slow grin spread over Dainis's face. "Rah!" he yelled, his hands darting out as if to chase the meat from his plate. "Rah, rah!"

Now, Dainis and Jānis join me outdoors. We leave the chickens to their rest and head down the back road, wearing orange vests and toques. After being confined to the house for much of the past week because Jānis was sick, we're desperate for an outing.

The boys romp and chatter ahead of me. They hurl snowballs at hemlock trunks, then at each other, then at me. I retaliate, and laughter fills the woods. Jānis nails me with a wad down my neck, and he and Dainis howl with amusement as I cringe and try to shake the snow away from where its cold moisture trickles down my chest. I press a handful of snow to Jānis's cheek and manage to shove a few small lumps down Dainis's sweater. More laughter fills the air, and our walk ends with me racing for the house. "Get her!" Dainis yells, and my sons' guffaws follow me through the snow.

November 28 (-13°C, overcast, moderate NW wind, 7:30 a.m.)

The predominant colour this morning is white. White piled on the car. White plastered against the coop. White weighing down spruce boughs and riding high against my rubber boots. I feed the chickens and cats, then the wild birds. Sunflower seeds spill black and shiny into the feeder, some falling onto the snow below. Chickadees are ready and quick, their black caps and grey-white bodies as winter-coloured as the shadowed tree trunks and snow-covered twigs at the yard's edge. Where is the brilliance of blue this morning? The blue jay. Of yellow and black? The evening grosbeak. They haven't arrived yet.

Wind whips branches and pushes diesel exhaust horizontally from our tractor. Yesterday, a winter storm jiggled power lines, closed highways, and left us with our truck and car stuck in the yard and a 200-metre expanse of wind-packed snow between our house and the shelter of the woods. Each winter we pay for the solitude we enjoy at Ravenhill, with no other houses in sight and an 800-metre access road.

Vilis rocks the truck forward and backward, then shovels snow from around it. I help, pushing and shovelling. Finally, we free the Ranger and move it out of the tractor's way. Then the old David Brown roars and lets out the reins on its horses while its chained tires hold firm in the slick snow. Two hours later, when the plowing is done, we free the car and Vilis drives belatedly to work through a white world. Every prize has its price, and price of winter life at Ravenhill is plowing.

November 29 (-14°C, clear, strong W wind, 4 p.m.)

It's too long since I last skied. Part of me comes alive as I kick and glide, revelling in wind-blown snow and pale magenta sky at sunset. Trees are weighted with snow—spruce boughs hang like arms pulled down by balls and chains, and low apple boughs etch grey curves that disappear in whiteness. Deer tracks punched in snow lead to tall grasses at the edge of lawn and garden, to fallen apples and to three broccoli plants I left growing in my garden. The broccoli stalks are shredded now, the leaves and florets gone, consumed. Winter is a hard time for deer, with their slender legs and small pointed hoofs. Deep snow steals their grace, leaving them awkward and floundering in their escape bounds.

Snow is a remarkable landscape-sculptor. It inherently changes the shape, texture, and density of the space immediately above the ground surface. No longer is that space occupied only by air and the stems of plants growing out of the soil, as it is in summer; rather, that space in winter is also occupied by varying depths of frozen precipitation—snow, or sometimes, ice.

Each year, terrestrial mammals that live year-round here in northern Nova Scotia face the striking change that occurs in their environment when winter snowfalls arrive. Some, like bears and chipmunks, hibernate and thus avoid having to search for food during cold snowy months. Snowshoe hares and lynxes have built-in snowshoes; for them snow is simply a different medium to move over. For meadow voles, snow is a boon that provides insulation from frigid temperatures and a protective screen from predators; the voles simply tunnel through it at the ground surface. But for our white-tailed deer, snow is an obstacle they must overcome by punching out well-used trails to favourite feeding areas. For the local deer, those feeding areas include our yard.

I snowplow down the front-road hill, exalting in the smooth glide of my skis. At the base of the hill, snow-laden alder branches droop into the freezing pond. Between pond and bridge, young birches and pincherries are bent into complete arches, with the trees' crowns of slender branches trapped in snow. I can't free them and can't get through them, so I ski up the hill and through our south meadows. If in this moment someone asked me to move back to the Yukon, I would say yes for no other reason than that at Kluane Lake I could ski from mid-October through April. Here in Nova Scotia, the snow might melt tomorrow.

It's dusk when I return to the house—clear white-snow dusk, clear blue-sky dusk. I plug in Christmas lights that follow the porch roof, then quickly unplug them. They seem wrong on an evening like this. I want only what's already here. Clear white dusk. Clear blue dusk.

December 2

Freezing rain fell in the night, sheeting our yard with glistening ice. We lost power, so now for fun the boys and I cook molasses-on-snow candy the way Laura Ingalls Wilder did when she was a young girl still living in the Little House in the Big Woods. To make the old-fashioned treat, we measure molasses and brown sugar into a heavy pot and cook the mixture on the wood stove until a little of the syrup dropped into a cup of cold water forms a hard ball. Then the boys fetch in pans of clean icy snow, and we pour the thick hot syrup onto the snow in squiggles, circles, and lines that harden in an instant and taste sharp and sweet in the warm light of wood fire and candles.

I don't ever recall seeing an ice storm when I was growing up in Alberta. I experienced my first such storm in Manitoba a few days after Vilis and I moved to that province's southwest corner in 1983 to begin his skunk research. For two weeks, we camped in an old van in Turtle Mountain Provincial Park. On our third night there, in early May, we woke in the night to the sound of rain plinking hard against the van. In the morning, we opened the van's rear doors and saw a world of glistening ice—ice coating the ground, trees, and van; ice hanging in long shining teeth from the van's bumpers and a picnic table. The storm lasted all day and at first enchanted us. We donned raincoats and rubber boots and explored the glittering world surrounding us. Later, chilled to the bone, we huddled in our sleeping bags in the van, trying to get warm and stay warm.

Here in Nova Scotia, ice storms, often accompanied by power outages, are simply part of life. The sudden silence, darkness, and cold initially unnerve us, but then our sense of self-sufficiency kicks in, bringing with it an air of adventure. We light candles, cook food on the wood stove, and, in doing so, catch glimpses into the lives of others who throughout history huddled close to fires while storms wailed beyond their shelters. These storms test and challenge us. So far, we haven't been found wanting.

December 8 (overcast with snow, NW wind, 8:45 a.m.)

Snow grunts under my skis as I ascend a gentle slope behind the house, picking my way through sprawling apple trees and upright spruces. I'm looking for snowshoe hares, for black-marble eyes and flicking white ears, but I see only "snow hares," rounded shapes created by snow sliding off evergreen branches.

Two winters ago, I often saw hares on our land. One of them was so silent and still amid a patch of dried grass, it nearly fooled me into thinking it was a snow hare. I watched the shape for half a minute, and it never moved. I slowly raised my binoculars, and the hare came into focus—the ring of brown around its black pupils, the grey fur blending with white in its ears, the way its wet body fur waved into tiny white deltas of guard hairs over rusty underfur.

Later that same winter, another hare allowed me to observe it from only three or four metres away. Its fur was still white, and the linings of its ears held a mauve tinge in the shadows of the young spruce trees where the "rabbit" fed. Pea-sized brown spheres filled with tiny wood shavings—the hare's droppings—lay in or near its tracks in the snow. That hare moved its ears constantly, one angling toward me, the other forward, and its whiskers twitched as it tested the air with its nose. It nibbled at a spruce twig on the snow and then grasped the twig in its mouth—needles and all. As the hare chewed, its entire face below its eyes bulged and bunched, and I could hear its teeth grinding.

Perhaps my lagomorph acquaintance sensed no malicious intent on my part because it balanced on its hind feet and stretched upward to clip another spruce twig. In that moment I was struck by the hare's body shape—rounded and plump when hunched against the snow, and lean and long, with those power-packed hind feet stretched out, when reaching for the twig. The rounded, plump shape is designed for conservation of body heat, and the lean, long shape, for speed. How many other facets are there to a snowshoe hare's life?

Here's a bit of science. The snowshoe hare is what ecologists call a keystone species, that is, a species that's integral to maintaining the diversity of a given ecosystem. Snowshoe hares are browsers. They feed mostly on shrub and tree twigs and are, in turn, hunted by terrestrial predators such as coyotes and lynxes (the snowshoe hare's arch enemy), and by aerial predators such as northern goshawks and great horned owls.

In the early half of the 1900s, wildlife scientists began to take an interest in the Hudson Bay Company's fur trading records, which showed a pronounced ten-year cycle in the number of snowshoe hares trapped over a span of more than a hundred years. During each cycle, hare populations increased to explosive levels and then crashed back to low levels before beginning to rise again.[41] Intriguingly, the Hudson Bay Company's records also showed an obvious ten-year cycle in the number of lynx pelts trapped, although the lynx population cycle lagged a year or so behind that of the snowshoe hare.[42] Since snowshoe hares are lynxes' primary food, it's easy to understand why lynx populations would shadow hare populations, but what caused the hare populations to rise and crash?

To answer that question, a large-scale, multi-university snowshoe hare research project headed up by Charlie Krebs, Tony Sinclair, Rudy Boonstra, and Stan Boutin began at the Arctic Institute of North America's Kluane Lake Research Station in 1986. This was the project that took Vilis and me to Mile 1054 Alaska Highway, where I adorned our cabin wall with a flower box of pansies.

That region of the Yukon is wild, majestic country with glacier-gouged lakes, rugged hills, and soaring mountains. The valleys are filled with boreal forest—prime habitat for snowshoe hares and their predators. Because the Alaska Highway runs through Shakwak Valley, researchers had easy access to sites where they could test factors that might determine what causes snowshoe hare numbers to rise and crash.

The hares' population crashes are sometimes dramatic. While living at Kluane Lake, Vilis and I visited an elderly Southern Tutchone woman who resided near the station. She told us of a winter when there were so many starving hares, they came right into her family's yard and, driven by hunger, devoured all the hay intended for her father's horses. At night, she could hear the hares chewing the hay. The following spring, streams were clogged with emaciated bodies of dead hares.

To determine the cause or causes of snowshoe hares' ten-year cycle, the researchers at Kluane Lake did a huge experiment that had four parts or treatments. For one treatment, a control, they did nothing. This treatment had three reps. For another treatment, they added food (Purina Rabbit Chow) to an area of forest to see if extra food would keep the local hare population from crashing. This treatment had two reps. For the third treatment, they built an electrified fence (Vilis and I helped with that) around a one-square-kilometre area of forest to test whether a hare population that was protected from predators would sustain itself and not crash. Aerial predators were excluded from part of the fenced area by stringing monofilament over it. As could be expected from the amount of work involved in setting up that treatment, it had only one rep. For the fourth treatment, the researchers added extra food to *and* excluded predators from a one-square-kilometre patch of forest to see if the two factors worked in synergy. That treatment also had one rep. The project finished last year, in 1996, after following a complete ten-year cycle. Experimental results from the four treatments showed that the amount of available food *and* predation *together* influenced snowshoe hare numbers.[43]

Here at Ravenhill, the back road is rough and icy, so I ski among towering hemlocks. Mounds of snow under the trees are pocked with craters gouged by cascades of icy snow freed from boughs by sun and wind last week, so my outing becomes a roller coaster ride with detours around deadfalls, raspberry canes, and beech and fir saplings. I wind my way back toward the house and link up with the road, where I notice three sets of fresh hoof prints, perhaps made by the doe and her fawns, on my ski tracks. From the sign, I can tell that the deer advanced and then turned hastily and retreated along the road before bounding into the woods. Had I been looking ahead rather than at the terrain under my skis, I might have seen the trio, their ears flicking forward, their tails lifting in alarm, their steps quickening.

Snowflakes tickle my nose and dust evergreens with wild powdered sugar. Our meadows, with their open spaces and smooth beckoning curves, are freedom to my skis. I push, glide, and suck some of that freedom up through my boots into my legs, torso, and mind. Now I can write.

December 13 (overcast, calm, 4:15 p.m.)

Late afternoon air is cool and fresh, holding none of the sweetness I leave behind in my kitchen, where peanut brittle cools on cookie sheets and Mexican orange drops harden on waxed paper. I ski the gentle slope behind the house, smelling only snow and tasting only winter. Loath to trade the coziness of gathering dusk for candies and cookies that must be bagged and boxed for travel to faraway family, I ski past the house to the south meadows. This isn't a dusk for bright lights and candy, it's a dusk for dreaming. And for remembering.

When Jānis was three, he inherited Dainis's outgrown cross-country skis, poles, and boots, but for the sake of sibling harmony Vilis and I didn't allow Jāni to try out his skis until Dainis had new equipment. Fresh snow—thick and wet—fell on a Wednesday, but it was Friday evening before Vilis brought the new skis, poles, and boots home. The next day, rain and above-zero temperatures melted the snow. "It's not fair!" the boys howled, so Vilis fastened the bindings onto Dainis's new skis, and the boys skied without poles around the loop inside our house. Striding, stumbling, and laughing off their disappointment, they repeatedly circled from living room to hobby room to study and back to living room.

After days of expectant waiting, we watched new snow drift down on the following Thursday afternoon. Huge wet flakes quickly covered the ground and accumulated to form a thick expanse of enticing snow. At dusk, my sons and I stepped onto it, and away Dainis went—no staying close by, no looking back. And away Jānis went after him, sometimes holding my hand, sometimes on his own, following his brother's trail wherever it went as golden light from the house filtered through the spruce shelterbelt onto the snow. Our outing ended when Vilis opened the kitchen's kissing window and called us in for supper. That dusk was like this one—heart-stretching with the beauty of life and snow. I turn homeward, tuck memories in my pocket, and ski.

December 17 (clear, S wind, 3 p.m.)

Parcels are in the mail, bound for Alberta and Ontario. Cookies and candies fill shiny metal tins on pantry shelves. Paper snowflakes adorn windows. Handcrafted wreaths encircle thick candles. A wild asymmetrical spruce with proud thrusting branches tolerates ornaments and brings forest air into our living room.

Now I'm escaping. Escaping from children, teaching, my house—and from Christmas. I seek the forest air outdoors in the company of trees, wind, snow, and sun. This is a day meant for walking, for soaking up unexpected warmth on a December afternoon, for looking into puddles and seeing the bottom of an ocean or the top of a sky. My legs' need for motion is like a hunger. They eat up the back road and stretch to swallow Balmoral Road. I look and listen, my mind racing ahead, dashing behind, zigzagging and cavorting.

On Matheson Brook Road, two ravens sun-bask atop roadside tamaracks near garbage strewn at the corner of a neighbour's driveway. I put two and two together and wonder what delicacies the ravens found amid plastic and paper refuse. On our front road, I follow what I think are porcupine tracks. The prints show broad five-toed forepaws and hind paws with an obvious heel. This story in snow is real and true and, like the ravens, is here when I need it. Matheson Brook is all black water and white snow edged with ice. Upstream from the beaver dam, ice on the pool under the bridge is silver and pebbled, serene and warty.

As I slog up the hill on crusted snow, every particle of my being rebels against going indoors. I want to remain in this world of snow and ice, ravens and porcupine tracks, and puddles that reflect the bottom of an ocean or the top of a sky. Motherhood, like all things worth pursuing, comes at a cost. Every prize has its price, and for me that price is a lack of time and freedom for my own needs and interests. I've never doubted that the prize is well worth its price, but today I rebel against paying it. This walk, this time alone to observe the world around me and have it wake my awareness is an absolute necessity. It takes me outside the cloistered bubble of family life. Here in the outdoors, I catch glimpses of the big picture, as though I'm looking into the puddle of my existence and seeing the reflection of God's ocean-bottom or the top of his sky. And I want more of this revelation, more opportunities to look deep into the puddle, my gaze led by porcupine tracks and ravens perched atop tamarack trees.

In his "Auguries of Innocence," William Blake wrote: "To see a World in a Grain of Sand/And a Heaven in a Wild Flower/Hold Infinity in the palm of your hand/And Eternity in an hour."[44] On this walk and others during the past eight months, I've caught glimpses of infinity and eternity. For infinitesimal moments, I've held them in my mind. I don't want this journey to end. I don't want to go in.

4

A Million Diamonds

December 27 (4°C, overcast, calm, 3:25 p.m.)

The chainsaw roars, a hatchet lifts, and white-and-purple sneakers dance. With a heavy snowfall warning in effect for tonight, Vilis is sawing as much of the woodpile into stove-box lengths as he can. Dainis uses a hatchet to knock logs free of ice, and Jānis plays on crusted snow beside the woodpile, his jacket a splash of red, his sneakered feet skittering.

Ice sheets the yard and back road, and a few patches of exposed gravel attest to rain that fell on Christmas night and to above-freezing temperatures yesterday and today. In the place of change at the edge of the woods, ice is replaced by thick snow the texture of rough sugar. In the forest I alternate between sliding over shiny ice, with my steps short and wary, and crunching through crusty snow. I slip once but catch myself before I fall. Observation is a cautious venture with ice under one's feet.

The woods are silent and gloomy. Air hangs damp and heavy under pale grey cloud. A beech leaf—oval and pointed, scallop-edged, copper and gold in hue—lies in the middle of the road, fallen or torn at long last from tenacious wooden fingers. A few steps farther along I encounter another beech leaf, and then another, but neither of these shares the perfection of the first. At the top of the back road, my sneakers claim no traction at all, so I carefully traverse the slick road to reach crusted snow at its edge.

Balmoral Road's pavement is like open sky to a tethered dove. Now I can look up at rumpled grey clouds shrouding the sky to the north, east, and west. A red squirrel chatters faintly from the back of the clear-cut to the south, but other than that, heavy silence hangs in the air. Matheson Brook is no more than a whisper flowing through a huge culvert near the intersection with Matheson Brook Road.

As soon as I step onto the gravel road, which is coated with ice and snow, my attention again shifts to my feet. A frayed shawl of sawdust shows where bent and broken limbs were sawed from roadside trees after the ice storm earlier this month. The sawdust gives way to dirty crystalline snow that shatters cacophonously when I walk on it. With each step, my sneakers crush and grind brownish frozen pedestals, causing them to crash against each other with a jarring racket. My passage is so loud that all other sound is dimmed or fragmented.

If I were to hunt in a world of such ear-cringing sound, my sight would need to become quicker and sharper. On the other hand, if I were to hunt in a world of slick, slippery ice, my hearing would need to become keener and more discerning because my eyes could afford only brief, darting glances away from my feet. Of course, over time I would change and adapt. I would learn to walk more quietly or less cautiously and to find my prey by sight or sound as the situation demanded. Or I would die.

A sparrow's "tsip" breaks into a moment of quiet in the clamor, and then the soaring hum of an airplane flying high above the clouds. On the front road, my feet punch deep holes in crusted snow, and I tread carefully across the slick bridge, its upstream edge slanted toward the water. I climb the hill at the edge of our toboggan run where packed snow is slippery but hard enough to support my weight. Tiny, toothed wire birch seeds are everywhere, blown down the slope from the copse of birches near the top of the road. Breath saws in my throat, but my feet have exciting stories to tell—of shining smoothness and crumbling towers, silence and cacophony—and of survival.

December 30 (10°C, cloud/clear, S wind, 12:15 p.m.)

On Christmas morning, I walked the back road to the woods before sunrise, the only sound the snow squeaking under my boots. The coop, meadow, and woods were silent, as if the whole world were waiting to again celebrate Christ's birth.

Today holds none of that snowy silence. The weather was wild when I rose from my bed. Rain and a south wind lashed the house, and the temperature soared to 10°C. Now, in late afternoon, the rain has ended, but the temperature remains warm. Stubborn sunshine pushes through thick grey clouds as I slip and edge my way along the back road. Mist hangs in that place of change at the edge of the woods. In the forest, a wind gust shakes raindrops from branches onto my head, and I step into soggy roadside snow to avoid a large puddle in the Cathedral.

Gone, at least temporarily, are my solo skiing outings and those with my family. Two days ago, as Vilis, the boys, and I skied over our land, a fierce wind blasted our faces with snow and filled in ski tracks behind us. We poked along in the meadows south of the house and scrambled through spruces and alders to the meadow beside Matheson Brook. We also skied on Christmas morning and then invited Blotchy and Charcoal indoors for a visit while the turkey roasted in the oven and we listened to a new nature-sounds CD.

That day I learned, yet again, that revelation is spontaneous, that out of nowhere something unexpected might shed brilliant light into my understanding. Charcoal, who had been casually strolling through the kitchen, tensed, his tail flicking. He stalked toward the living room and looked up frequently while the CD played a common yellowthroat loudly singing its "witchy-witchity" song. A blue jay and several white-throated sparrows sang in the background. The previous track had played sounds of the Pacific shore, which had elicited no response whatsoever from Charcoal, yet now he was stalking.

A light bulb went on in my brain. *He knows our birds.* This cat was a birder in the most elemental way—as a hunter. The yellowthroat, blue jay, and white-throated sparrows he heard on the CD were species that inhabited Ravenhill, and he recognized their songs. Charcoal sat on his haunches and stared up at the stereo until the yellowthroat's song faded. That was five days ago, and now, as I walk the back road, my comprehension still trembles and cheers.

There are times when even the warmth of a glass-fronted wood stove can't match the warmth of family. This is such a time. It's evening, and I'm reading in the kitchen, basking in the wood stove's radiance. In the living room, Vilis and Dainis, covered with a fuzzy blanket, read and giggle on the couch. Across the coffee table from them, Jānis rips through pieces at the piano, his exuberant playing and exhilarated commentary prompting most of his father's and brother's chortles. Dainis visits me in the kitchen and tells me with hushed excitement that he's sucking a candy that tastes "like mint and root beer, like spring and fresh air." It's wintergreen.

"Hey, Mom!" Jānis calls. "I finished *Piano Plus: One.* Now I'm doing *Piano Plus: Two*!" He's already played for nearly two hours, with wild abandon and a total lack of regard for the finer points of timing. "Uh-oh!" His voice is full of delicious apprehension. "The next one's "Waltzing Matilda!""

"*Ooooh!*" Dainis and Vilis coo in expectant unison, and all three of my men break into silly laughter as Jānis tackles the piece with flying fingers.

At 10:30 p.m., I close my book and wander into the living room, ready to climb the stairs to bed. Not my guys. They're still full of chuckles, piano, and late-night Christmas holiday magic. And so am I.

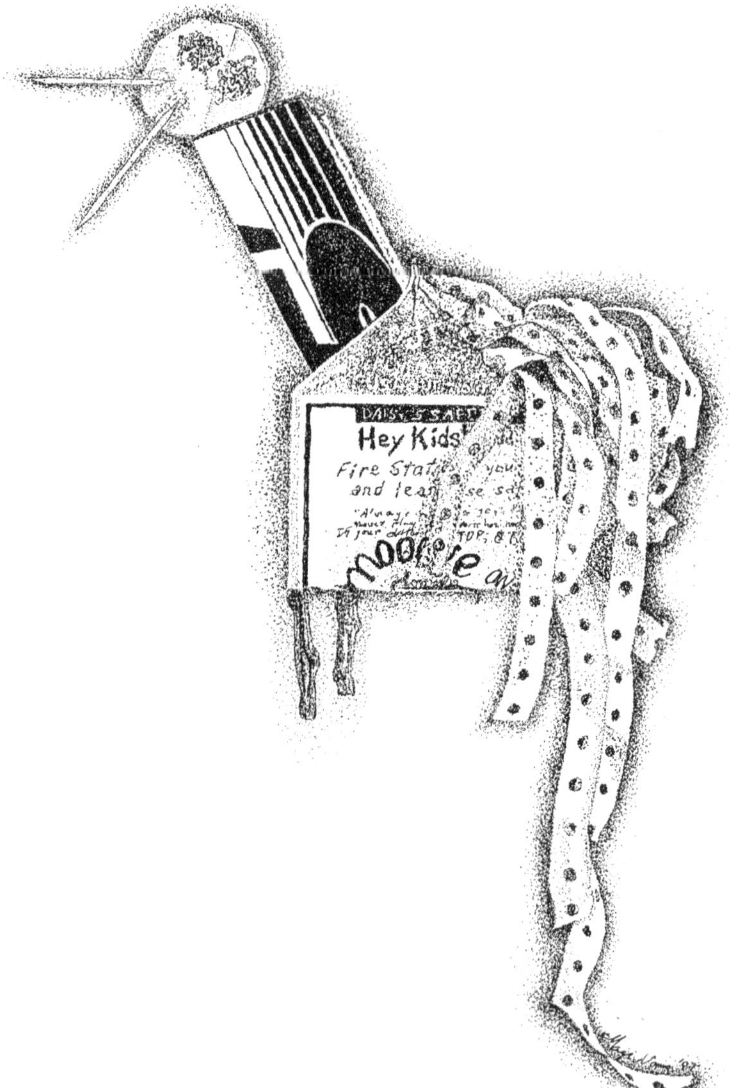

January 4, 1998 (Clear. Windy. 2 p.m.)

A WINTER PICNIC IN 42 WORDS

Blue sky. Evergreens. Walk-on snow.

Fire.

Wieners and challah. Hot chocolate. Last Christmas cookies.

Sun.

Wind-hidden. Snow naps. Playtime.

Boys.

Tree climbing. Branch sliding. Mother hiding eyes.

Me.

Toes cold. Fire fading. Crater melted.

Time.

Kick snow. Lift packs. Smile home.

January 10 (0°C, overcast, calm, 10 a.m.)

This morning, ice coats every grass stem, tree trunk, branch, twig, and needle. The world on my doorstep is sheathed in silver. Birches along the front road are bent curves lined with ice. Near the backroad woods, ice-laden alders droop onto the snow, some branches encased in frozen water a centimetre thick. What is the weight of this burden? The very advantage of branching to finer and finer levels to increase leaf area and, thus, photosynthetic capacity, has backfired. Each alder nutlet, catkin, and twig is encased in an icy sleeve whose weight it cannot bear.

In the woods, hemlock boughs slump beneath loads of crusted snow. Ice droplets hang like pendants from needles. If today were a sunny day, this ice-world would be a realm of sparkling crystal and small rainbows leaping from every branch. Surprisingly, the glitter storm didn't bring down any trees on the back road, no barricades of thick wood that won't budge without the convincing bite of a chainsaw.

Slush and ice cover Balmoral Road, with only a single bare strip worn through to the pavement at the double yellow line. I hear no bird or squirrel; rather, the constant drip of melting ice, the noisy rush of a small stream, and the occasional whirring thud created by ice sliding off a tree and smashing onto crusty snow below.

Matheson Brook Road is a white expanse with a lone silver-grey vehicle track curving down its centre. Here, the snow is surprisingly resilient. It bounces slightly with each step I take and then gently sinks into compact sneaker-shaped prints. A goldfinch calls, "see-*ee*," as though asking me to take notice. "Yes," I assure it, "I see."

I break the ice from our zebra mailbox, flip its ear up, collect mail, and then tackle the front road. Ice-sheathed branches tinkle like glass curtains as I brush them aside. Beech leaves flame wet and tawny against dark spruce trunks. I avoid holes in the bridge and, as I jump to terra firma on the far side, spot fresh beaver tracks—broad webbed hind feet—and a chewed branch at the edge of the pool under the bridge. Downstream, water pours through a gap punched in the beaver dam by the rainstorm that whipped our land in late December.

I see no obvious recent repairs. Perhaps the beaver who left the fresh tracks and chewed branch is intent only on escaping starvation.

Near the top of the hill, I walk beneath a wire birch's perfect arch, the young tree bowed by its ice load. In his poem, "Birches," Robert Frost penned verses describing how ice storms bend birches into permanent arches in a way that boys swinging on them never do. He was right, but I think this birch will recover. Its twigs aren't trapped in the crusted snow, as are twigs of many of the young birches growing beside Matheson Brook Road. On a sunny breezy day yet to come, the ice sleeves sheathing this birch's twigs will tinkle like chimes in the wind, smash against each other, and break into shimmering lengths of clear split tube that lie like broken beads on the snow. "Such heaps of broken glass to sweep away/You'd think the inner dome of heaven had fallen," wrote Frost.[45] Yes, and then this birch will again lift its boughs to the sky.

January 11 (clear/cloud, strong SW wind, 9 a.m.)

A band of grey cloud races east across the southern horizon, with mares' tails flying high and white above it. Much ice has fallen from trees, yet much more remains, snapping and crackling around me as I walk the back road. Shafts of sunlight playing over ice-coated shrubs set them shimmering. Such splendor amid such hardship.

The storm cut a wide swath through eastern Canada. Here in Nova Scotia, we caught only its fringe. Our brief power outages and scattered broken boughs have been minor inconveniences compared with the devastation wreaked upon central Canada. For almost a week, the glitter storm battered southern Quebec and eastern Ontario with its full force. Now, finally, the tempest is over there, too, but it left human deaths, downed hydro towers, and countless broken trees in its wake.

I like the smell of the air today—fresh, cool, drier, promising sun. On Matheson Brook Road, the ghosts of my footsteps from yesterday's walk remain as grey shadows beneath the white shoulder of this freshly plowed road. Our zebra mailbox, its hinged mouth open, lies sprawled on the snow in the ditch, a victim of the plow. Fortunately, the board onto which it was screwed, and not the zebra itself, is what broke.

I take the zebra with me and today walk our front road atop crusted snow that yesterday cut into my ankles when I punched through it. Beads of ice lie strewn across my path. The "inner dome of heaven" has fallen. Beside me, the brook is a black curve between white banks. I hear the rush of water through the dam and look for fresh beaver sign but am disappointed. The chewed stick I spotted yesterday is encased in ice flakes, and no new tracks print the pool's edge. After crossing the bridge, I spot hare and squirrel tracks in frost on the road, and a flock of common redpolls chittering in a wire birch part way up the hill. The tiny finches tug open buds and catkins, spilling their refuse in brown flakes onto the snow.

Every prize has its price. As I warm up in the kitchen after my walk, my sons and I observe a shocking example of this maxim. Today, the prize was sunflower seeds, and the price, a squirrel's life.

Three days ago, Dainis, Jānis, and I first noticed one and then two red squirrels skittering about beneath the bird feeder in the front yard, stuffing black oil sunflower seeds into their mouths. Our cats also noticed. Blotchy made a half-hearted rush for one squirrel but gave up when it sprinted for a tree. Charcoal's hunting efforts were far from half-hearted. He crouched, stalked, and sprinted, his charges sending the squirrels racing in desperate headlong flight from the reach of his paws.

In the kitchen, the boys and I laughed at the zigzag changes of direction, the slipping cat paws, the mad dashes for the elm tree that ended with the squirrels chattering angrily from high in its branches, and Charcoal twitching his tail below. As soon as Charcoal abandoned his hunt, the squirrels resumed their dangerous pursuit of the sunflower-seed prize.

Today, the story has a different ending. A squirrel hangs limply from Charcoal's jaws, its silvery-grey belly fur bright against his black-and-tawny chest. There are no giggles in the kitchen, only the shocked realization of the suddenness with which death can strike. A mad dash. A wrong turn. An apple tree too far.

The prize. The price. Always.

January 12 (clear, calm, late afternoon)

This is the time of year when life closes in around me, sets its weight upon my shoulders, and spills darkness into my soul. I push against the seasonal depression, struggle out from under it, and send myself beyond my mind into the world of snow. I see where crow tracks appeared and disappeared with the brush of a wing, and where a deer mouse's small hops left sign for all to see that it ran from the base of a pincherry tree to the shelter of two spruces. I find comfort in pieces of old apple pecked at and tossed away by the crow, and in drag marks left by the mouse's tail. Other lives, other stories, free me from my own.

In "Birches," Frost wrote of being a "swinger of birches," and of becoming "weary of considerations" when "life is too much like a pathless wood." Maybe I need a birch to swing on in this doldrum period when January's long dark nights make my life seem like a pathless wood. Maybe I need to climb high in that birch, work my way upward among its branches, and then leap out while holding on to an uppermost branch, swinging in the air before I let go and seek the earth with my feet. That would be freedom. Or maybe, as Joni Mitchell sang, I need a river to skate away on, a river that takes me gliding away from the confusion of who I might be and brings me back to the woman I truly am.

Who can say where we find ourselves? In the soil? In the air? In shafts of sunlight that strike ice-coated shrubs and set them shimmering? In mist that hangs in that place of change at the edge of the woods? Who is this woman who nine months ago set out on a walk of freedom and discovery along an old wood road beside Matheson Brook? In these winter days when light is short and darkness clouds my spirit, I can't really see her.

January 13 (-14°C, clear, calm, 6:40 a.m.)

Before dawn, a blue world greets me at the edge of the porch. Snow. Sky. Moon. Stars. I'm draped in their dark light, borne upon their bare beauty between house and coop.

Flick a switch. Step into a brown world. Wood. Feedbag. Chickens. Incandescent lights startle and pull me instantly into this other world. Such a dichotomy—one world, blue and cold; the other, brown and warm.

I shut the coop door, revel in the quiet blueness that envelopes me, and give myself back to the night, to the moon, the snow, the sky. In their cold beauty I'm at peace. I know that beneath the vastness of the cosmos above me I'm small, a speck moving on the face of one planet in a galaxy among galaxies. Yet, this very vastness, with its cold blue-and-white beauty, feeds my soul. It comforts my yearning heart that doesn't even know for what it yearns. Yes, in the grand scheme of things, I'm insignificant, but I'm also part of something greater, something far beyond what my mind and senses can grasp. On the sixth day God looked at all he had made and said it was very good. I look up at the sky and say Creation is exquisite. This morning, before the rooster even crows and sunrise brings light to the world, I have hope.

January 23 (-19°C, clear, light N breeze, 8 a.m.)

The porch snaps and grunts under my boots. In the yard, dry snow squeaks like rubbed Styrofoam when I step on it. A bitter north wind whips my face. In the middle of a cold snap, I've returned to this world.

A week ago, I woke with strangers in my mind, fictional characters who spoke gritty dialogue in a dystopian future. They stole me away, catching me up in their storm of truth and passion. I was their scribe. With pen and paper (somehow, a computer didn't work for this), I recorded their words as scene after scene played through to the story's climax and resolution. That the fiction was a drama surprised me. The last time I wrote a play was in eighth grade, when two friends and I won a cake for our comic performance of my skit.

Now, as I walk the back road, the dream-play's characters—so familiar to me after spending seven days in their world—float along with me. I hear their voices among those of chickadees and see their faces in trees' shadows. It takes a long time to climb down the ladder from the garden of fiction. And that garden is boundless. At the end of a bout of fiction writing, I'm always amazed by what came out of my brain—dialogue, action, riveting conflict, stakes so high they leave me tense and emotionally revving. I become captive to the world I created, bound to a story that unfolds before my eyes, often surprising me and sometimes shocking me with its twists and turns. I become my characters, speaking with their voices, reacting with their senses, facing their hardships, cherishing their joys. At the story's end, it's hard to let them go, to leave them and continue on in reality without them. So, today I walk the back road accompanied by friends that don't exist, that never existed except in the rich soil of my imagination.

I reach the Cathedral, breathe in deeply of reality, and keep on walking.

January 25 (5°C, rain, strong SW wind, 9:30 a.m.)

All night, a southwest wind howled and slapped rain against our windows. This morning, according to radio reports, Truro's low-lying streets are flooded. Here at Ravenhill, water pours over our front road between the pond and bridge. Ice jams are responsible for both floods—Truro's, on the Salmon River; ours, at our bridge.

At the bottom of the hill, Vilis and I tread carefully through the rushing floodwater, the front road's icy surface uneven underfoot. Water rushes past only an inch or two below the tops of our rubber boots. Then we're on higher ground next to the bridge, which is almost invisible beneath slanted ice plates crushed together on top of the deck and piled at its upstream edge. Vilis retraces his steps through the flood and carries first Jānis and then Dainis to where I stand staring at the bridge.

"I think it'll be gone this spring," Jānis says, and we all nod.

Yet how many other springs have passed when flood waters poured over the front road and spit out beyond the ice jammed atop this bridge? And still it remained, ever weaker, ever more rotten, ever more sunken down to meet its fate of ice and water. I'll rejoice when we have a new bridge later this year and can once again drive the front road and walk across Matheson Brook on sound wood. But for now, we return through the flood on slick, rippled ice, with the sound of the freshet rushing against our ears.

January 30 (1°C, overcast, calm, 6 p.m.)

The boys and I are walking in the night. Dusk is newly fallen blue and thick onto clean white snow. I soak up the fields, the sky, the golden light spilling from our house and coop windows.

Not my sons. They belch and guffaw and hurl wet snow at each other, stopping a half-dozen times to empty snow from their boots. They're giddy, downright *giddy*, and I think, why not? Why not be giddy when life is filled with laughter, when you have a brother to throw snow at, and when you can rip out a good belch or two…or five or six? This is what Vilis will see when he drives home from work on the back road—this snow, this warm light spilling. I wonder, can he hear the laughter?

Four days ago, I heard black-capped chickadees whistle their "*cheese*-burger" mating call for the first time this year. It was as though a switch flicked. My mind leapt ahead to spring. I thought of gardens and ordering seeds, of spring cleaning and tulips. And for the first time in months, I thought of black flies, of how they swarm around us like a black cloud, of how their bites sting and itch, and of how we sometimes run from them, yelling, "I hate these flies!" Tonight, however, here on the snowy back road, the only yelling is in play, the only running is in fun, and black flies and the frustrations they bring are as far away as the moon we can't even see.

February 3 (-6°C, cloudy, calm, 8:30 a.m.)

Before sunrise, eastern clouds are like an expanse of coral-coloured fluff. I fill the bird feeder, its wooden walls rough with frost thorns. Then I lure Dainis and Jānis out of bed with French toast, and afterward we ski through the south meadows in a big loop and up and down the gentle slope behind the house. "Track!" we yell in warning as we push off from the top of the slope, double poling to pick up speed and then gliding as far as our skis take us.

Finally, with red cheeks and happy hearts, we come indoors to hit the books. Math speed drills. A language test for Dainis. Piano theory for Jānis. Piano practice. Compiling data and solving mathematical word problems for Dainis. Sentence exercises for Jānis. Books about stringed and brass instruments for us all.

The morning wings away. We rush through lunch, drive to Tatamagouche to do banking math, drive to Wallace to attend piano lessons, back to Tatamagouche to attend figure skating lessons, and then home, where I rush to put supper on the table. Yet, even with all the hurrying, part of me rested today on a cloud blanket at dawn, pricked a finger on frost thorns, and glided endlessly down a gentle hill, my skis carrying me on toward eternity.

February 8 (-10°C, mostly clear, calm, 8:30 a.m.)

Frost sparkles on every surface, luring me outdoors and into this ice-and-snow-dazzled world. Blue jays scream and fly from an apple tree, leaving stillness and silence in their wake. Sunlight tosses glitter onto our south meadows and paints gold over hemlock trunks in the forest alongside the back road.

I turn onto Balmoral Road and spot two people and two dogs far ahead of me. Abruptly, the road is no longer mine, the aloneness in which I wrap myself on these walks stretching and ripping to include others. My word images fade, and my ties to the natural world loosen. The forest that only seconds ago seemed so close, now retreats. Veiled by the presence of other humans, my eyes no longer seem to see it. This reaction startles me, but then I recall a time when Vilis and I canoed a wilderness lake for its peace and perspective only to be disappointed when we encountered other canoeists sharing the same lake. Our sense of aloneness and closeness with nature vanished.

The two people ahead of me walk only as far as Matheson Brook Road before turning back. As we approach each other, I realize they're strangers rather than the neighbours I thought they might be. The two dogs—one a handsome Siberian husky, the other a golden retriever—strain at their leashes. I greet their owners with a cheerful good morning, and we comment on the bright, invigorating day.

Then they're behind me, and the road ahead is mine once more. Cold bites my cheeks and toes, telling me my flesh is alive, this day is real, and I'm living it. My wits sharpen. My sight quickens. Curiosity arises, and word images play.

On Matheson Brook Road, advertising fliers are rolled, printed straw protruding from our zebra mailbox's gaping mouth. The last stretch of my loop, our front road, is hell. My sneakers punch deep holes in icy snow with every step. Beside me, the brook is jagged ice disguised with innocent snow. Past the bridge, as I pant up the hill, snow flashes a million diamonds around me.

February 10 (-18°C, clear, light SE breeze, 6:35 a.m.)

Stars stud early morning's blue darkness, one of them, a brilliant solitaire gleaming in the southwest. The Maasai people of East Africa tell the story of Kileken, the orphan boy who comes each morning to milk cows and lead them to pasture, and who returns with them each evening to milk again. It's Kileken I see now in the southeast sky—the morning star, the evening star, Venus.

I walk the back road and feel the deep, pregnant silence of winter before dawn. The squeak of snow beneath my boots instantly transports me ten years into my past. The ploughed snow under my feet is a Yukon airstrip. Trucks rumbling in the distance travel the Alaska Highway. Golden light shines from a log cabin while Sheep Mountain rises tall and jagged beyond Kluane Lake, there in my imagination, in my memory.

Jumbled thoughts. Distant places. The cold is here and now, stinging my face and grasping each twig and trunk with rough whitened hands. I return to the yard, feed chickens and pour sunflower seeds for wild birds. Like Kileken, I, too, have chores.

February 16 (-14°C, cloud/clear, light N wind, 8:25 a.m.)

Black-capped chickadees gossip and flit in the hemlocks near the top of the back road, seeking seeds and hidden insects to fuel their metabolic fires on a cold winter's day. That such small birds survive year-round not only here in Nova Scotia but across North America forever astounds and comforts me. With their smart black hoods and neat black bibs, black-capped chickadees are like friends I wave goodbye to in one part of this vast continent only to find them welcoming me when I move to another. I'm in awe of raptors, thrilled with warblers, and moved by skeins of geese honking overhead, but I love chickadees. They embody life's energy crammed into tiny exuberant packages.

Balmoral Road stretches before me, then Matheson Brook Road, and then our front road. I have only one thought—today I walk with joy, the joy of moving freely, bathing my soul in sun and wind. Summer lies in wait, sultry and lush, but for now I feel the clean biting kiss of winter.

In late afternoon, my sons and I head down the hill to play hockey. Our rink is the little pond near the brook, its pebbled surface humped and split with long deep cracks that show an impressive ice thickness. Because the surface is so rough, we wear boots rather than skates. I'm one team, and Dainis and Jānis are the other. The boys' goal, marked out with sticks, is off to one side near the broad end of the pond, avoiding a seepage zone's thin ice. My goal is near a narrow neck that connects the pond and brook at high water.

Jānis and I face off at centre ice, and Dainis drops the puck, which is a square of wood painted neon orange. Sticks clash. "Get it!" Dainis shouts. "Uh-oh! Here she comes!" Jānis yells, laughing. I shout and laugh, too. We stickhandle the puck, steal it from each other, and shoot it at each other's goal. Jānis swings like a golfer and misses the puck again and again because his stick's blade has a gap in it. I score, and the boys tie it up, but then I score two more goals.

The puck gets caught in cracks in the ice, lost in the snow, hung up in twigs of alder branches trapped in the frozen pond. We poke it out and fight over it, scoring more goals. The boys learn to pass behind me and to shove the puck ahead of themselves on breakaways. I pull ahead 5 to 4, then 6 to 4, but my sons fight back and tie the score just as the light becomes too dim to see the goal posts. Darkness settles gently around us as we—flushed with hockey, winter, and life—climb the hill.

February 18 (4°C, clear, E wind, 3:45 p.m.)

Today I seek the back road like a child's hand seeks its mother's. I need animal tracks and silence. I need sun and shadows and solitude.

Beside Balmoral Road, beech leaves rustle in a breeze. Small streams chuckle. I imagine hearing juncos' mating trills and white-throated sparrows' three-note melodies. On days like today, when the sun climbs high, it's easy to think that February leads straight into April and forget that March comes between. An east wind smelling of spring adds to the illusion, yet I know we may still have storms aplenty. Nova Scotia's late winter weather is wildly variable. I've visited schools for writing workshops in mid-April and driven into the teeth of wind and snow hours from home. Even later, early May seems to be a magnet for storms. In two of the six springs we've spent at Ravenhill, snow has fallen on May 6.

At the other end of the spectrum, temperatures have sometimes soared to near 30°C in May; more than once, spring has erupted with breathtaking speed. During our initial spring at Ravenhill, in 1992, the first truly warm day was May 8. That evening, while year-old Jāni slept, Vilis, Dainis, I, and Steve and Anne (two biologist friends on leave from work in Nairobi) saw an amazing sight. Patches of dry grass four or five metres ahead of us, at the edge of the yard, seemed to twist and writhe as we walked toward them. No matter which way we walked, the same thing happened.

We wondered if we were seeing things. Then we realized that we *were* seeing things—earthworms. Gobsmacking numbers of earthworms wriggled and squirmed into their holes as the vibrations of our footsteps warned them of our approach. I'd never seen so many earthworms, nor had any of us. There must have been thousands. As the oligochaetes writhed, they nudged grass stems, making the earth's surface shiver as if it were dancing a little spring dance. Dainis, just turned four, ran ahead, laughing as he bent to touch the earthworms and hasten their escape.

Now, on our front road, I hear black-capped chickadee whistles and think not about earthworms, but about the salamanders living on the steep, damp slope beside me. Like the rest of Nova Scotia's amphibians, eastern red-backed salamanders, yellow-spotted salamanders, and blue-spotted salamanders hibernate during the winter. In late autumn they seek out protected hibernacula deep below the soil surface—perhaps, a rodent burrow; perhaps, a crevice created by a tree's root. There they enter a state of dormancy and remain dormant until the warming air of spring lures them forth from their shelters. Today's warm east wind fibs that spring might arrive tomorrow.

Downstream from our bridge, water glug-glugs out from under ice and swirls away, brown and curled. The sun swallows me as I climb the hill by stepping into and over old footprints and even older parallel tracks, the legacies of walks, skiing, tobogganing, and playing hockey. Is anything else as clean for spirit and clothes as snow play? Is anything else as invigorating? Winter need not hurry away.

February 23 (-6°C, clear, calm, 7 a.m.)

Two days ago, evening grosbeaks greeted me in our yard, and a raccoon left its hand-like tracks neatly printed in frost on our porch's front steps. Crows, jays, and a raven called, and their voices, added to those of the grosbeaks, created a sense of urgency and busyness, as though the birds were discussing the amount of grass exposed by melting snow and the rush of spring this portends. Anxious for the new season to arrive, Dainis and Jānis fidgeted through their morning schoolwork. Their years are still so few that each season stretches beyond their patience. For the past week or more, I've been content to wait for spring because winter has brought gifts of sunshine and more snow than ice.

Today, however, my seasonal darkness descends, and I walk the back road with shoulders stooped from the weight of restless sleep and worries. I don't even reach the forest; instead, I'm lured from the road by twittering sounds and different voices in a young spruce tree in that place of change at the edge of the woods. Small grey bodies flit about in the spruce, one sporting a rich brown cap, another, a startling black-yellow-orange-striped crown, and a flock of others, white tail edges. My birder's brain clicks into identification mode. Boreal chickadee. Golden-crowned kinglet. Dark-eyed juncos. The tiny birds bring a flurry of action and sound to one small tree and gift me with a sighting of two species I hadn't yet seen this winter, the boreal chickadee and golden-crowned kinglet. Some of the weight lifts from my shoulders, and I know yet again that God sends me signs when I need them.

The dichotomy of my response to winter isn't lost on me. On one hand, I treasure this season's freshness and stark beauty; on the other, I dread the burden of its long periods of darkness. At Kluane Lake in the Yukon, the hours of winter darkness stretched even longer. For an entire month, the sun disappeared behind the mountains. Dawn stretched out for hours, as did dusk. Much of daily life was enacted in twilight, a twilight that became mirrored in my emotions. I never had suicidal thoughts, but for the first time in my life I understood how it was possible to have such thoughts. Even the majesty of the mountains and the entrancing wilderness surrounding the research station weren't enough to lift the burden of the North's winter darkness. Only the sun had that power. The first glimpse of its golden radiance spilling over the peak of Outpost Mountain in early January offered a beacon of hope. Today, my beacon of hope is twittering, flitting songbirds in a young spruce tree and the knowledge that the spring equinox is only twenty-five days away.

February 25 (6°C, overcast with rain, E wind, 3:15 p.m.)

Steady rain taps my baseball cap. Chickadees flock to exposed soil under the feeder in our front yard. Down the hill, Matheson Brook roars, its belly swollen and overflowing with rain that began in the night.

In the forest along the back road, mist hangs among the trees. Wind gusts knock heavy drops from trees into puddles and small rivulets of the kind that Jānis calls trickle-streams, the miniature waterways running in narrow beds carved in ice by car and truck tires. Hemlock trunks are rich with greys, greens, and black, the snow beneath them smudged and littered with twigs, cones, and needles. Here and there, bare ground rises in gentle swells capped by mosses, the bryophytes' texture, lush and frilly, their brilliant green, the verdant colour of another season. Winter's grip is easing, its cold breath warming, its icy touch melting. As I retrace my steps through the woods and meadows, rain envelopes me in the promise of spring.

March 1 (9°C, overcast, calm, 2:50 p.m.)

Grape hyacinths push clusters of slim green fingers into mild air. Melting snow reveals pansies and lavender, the pansies' flowers still vibrant purple and yellow, as if frozen in time.

"Spring is here twenty days early!" Dainis shouts. He and Jānis are romping on the chicken coop's shallowly slanted roof. Beside the coop, Vilis splits wood in his shirt sleeves. More than half of the six cords of firewood we bought in October remain, a testament to our moderate, although snowy, winter.

Green grass blades thrust through our lawn's worn yellow-brown carpet, and much snow has already peeled back from the south meadows. The back road is bare as far as the woods, with mud in its ruts. In the forest, snow still blankets most of the ground, the air above it hung with cold mist. On Balmoral Road, a woolly bear caterpillar, the larval stage of the Isabella tiger moth, crawls across the pavement's verge, its bands of reddish-brown and black bristles undulating like waves. Small tributaries of Matheson Brook that barely whispered in early winter now rush with sound and water. We're climbing the sine wave curve of spring.

In mathematics, a sine wave curve describes a regular, periodic pattern of increase and decrease, such as the pattern of increasing and decreasing day length. The curve goes up and down and up again, like the path a long ribbon would take if smoothly flicked up and down. A ribbon moves quickly when flicked down and then slows as it approaches the bottom of its path. An upward flick has it rising with increasing quickness, only to slow again as it approaches the top of the upward path. The ribbon moves fastest halfway up its rising path and halfway down its falling path.

In the same way, day length increases in ever greater leaps after winter solstice until the rate of increase reaches its maximum halfway up the annual sine wave curve. This point is the spring or vernal equinox when day and night are equal. After the spring equinox, days continue to lengthen, but in ever smaller leaps until the increase halts at summer solstice, the top of the ribbon's path. After summer solstice, day length shortens, at first slowly and then in ever greater leaps until the fall or autumnal equinox when again day and night are equal. After the autumnal equinox, day length decreases further, but in smaller increments until the lowest point of the ribbon's path, winter solstice. Then the entire flick-the-ribbon-up-and-down pattern repeats itself. There are times when I'm enchanted by mathematics. Spring is one of them.

Matheson Brook Road is gravel, mud, and potholes. Dirty snow in the ditches still traps branches of young birch trees bowed by January's ice storm. Beside our front road, Matheson Brook runs wide and bronze, with stranded ice plates clasping alders' ankles on its banks. Other ice plates melt on our bridge's sagging, rotten deck, which now has even more holes and is coated with a thin layer of silt deposited by the flood in late January. On the hill, I hop from grass clump to grass clump, moisture seeping through the sides of my sneakers, the grey sky like a rock on my shoulders.

March 9 (0°C, overcast with ice pellets, strong S wind, 9 a.m.)

Ice pellets pop off my clothes and glasses like miniature white BBs. Snow lies thick and clean in the meadows and woods. How much snow will remain after the rain forecast for today and tomorrow?

Winter returned six days ago on the wings of a northwest wind that dumped heavy wet snow onto northern Nova Scotia, Prince Edward Island, and parts of New Brunswick. Dainis and Jānis, who eight days ago reveled in spring-like temperatures, nonetheless joyfully built new snow forts and peppered each other with snowballs. And yesterday morning, Vilis, the boys, and I pulled loaded sleds home from our camping site near the Amphitheatre, revelling in the vitalizing afterglow of a winter night outdoors. Then we skied in T-shirts and played fox and goose. Our entire day was one last fling with snow, one last laughing, loving embrace of winter.

All around me now, a strong south wind speaks among the backroad hemlocks, telling a story of distant lands from whence the rain will come. Snow tumbles onto me, shoved from a bough by this wind that seems intent on shaking the forest into awareness of its strength and the change it will bring. On Balmoral Road, a blast of air throws ice pellets in disarray across the pavement, whips snow into my face, and mutes a raven's hoarse call. Today, the wind is a sky lion roaring, and all other voices bow before it.

On Matheson Brook Road, the roaring lion prowls high in the treetops, whipping spruce and pine branches this way and that in a disjointed dance, their needles bobbing like flotsam on a storm-tossed sea. On our front road, the lion's voice yields to the voice of Matheson Brook, rippling and strong, its power already building until sometime in the night its roar will rise and rival that of the sky lion brandishing all its strength. Our canoe, upturned upon a grassy hummock on the side of the hill, is well above the reach of the coming rain and the flood it might birth.

In the yard, evening grosbeaks in the young elm carry on a passerine conversation, their calls clear and concise, as though they were humans who spoke only in words of one ringing syllable. When I try to close the house door, the wind shoves it open, another reminder of the storm it will bring.

March 10 (11°C, heavy rain, strong SW wind, 6:40 a.m.)

I was wrong. All night the sky lion roared, its fury shaking the darkness, its batting paws making our old house shudder. It was as if the lion were seeking me out, seeking out the one who dared to say the brook could rival it. In the night, I didn't even hear the brook. I heard only the wind.

In our yard, water runs downhill under my rubber boots. Wind lashes me with rain. All the snow is melted. Overnight, thirty centimetres—a foot—of snow had disappeared.

I walk down the front-road hill, wondering if our sagging, decrepit bridge is still in place. Amazingly, it is. The brook runs fast and strong but quieter than I expected, its muddy water rushing in a torrent under the bridge deck and lapping at its upstream edge.

Back up the hill, mist wafts from thin patches of snow in the parkland. Behind the house, my garden is mud. Ripped loose by the wind, a section of the coop's asphalt roll roofing is a black curve against the sky. And still, the sky lion roars.

March 12 (-7°C, mostly cloudy, calm, 6:20 a.m.)

The bridge is gone. Late yesterday afternoon, the boys and I headed for the mailbox only to find our passage across the brook halted by a brown torrent. The bridge deck and most of the cribbing had been torn from the brook's banks, and only a few of the original logs remained, partially protruding from the turbulent water or caught in alders downstream. Where did the rest of the bridge end up? When did the brook finally tear it away from the banks?

The morning after the sky lion's night-long roar, the sagging structure was still in place, with water lapping at its upstream edge. That afternoon, Dainis, Jānis, and I drove to Tatamagouche and then to Wallace, and every brook and river we saw raced with high water, and low farm fields lay like lakes adjoining the rivers, with floodwater lapping at road edges and flooding driveways. Perhaps it was then that rushing Matheson

Brook, swollen with rain that fell throughout the day, swept away our old bridge.

Today, an immense grey cloud pushes southward out of the north like thick soft smoke spreading through a room. Fine snowflakes drift onto a dusting of snow that covered the ground last evening. I walk the back road through a silent world. No bird song or squirrel chatter. In the place of change at the edge of the woods, I see how snowshoe hares use this edge, this boundary between one habitat and another, as their highway. Their tracks are plain on the road snow, coming from the forest edge on one side of the road and going into the forest edge on the other side. Do they check for predators before crossing? A moving animal would be highly visible in this long narrow clearing, and hares have many predators.

When I reach Balmoral Road, I turn right rather than left. With the bridge gone, I can't walk my usual loop. One door has closed, but another stands open.

It's good to be up early again and to be part of morning's birth. Winter, with its long nights, steals away my rising energy. Spring brings it back on a platter of birdsong and brightening skies. Summer sees my rising energy at its greatest, when the cool freshness of dawn is a welcome reprieve from black flies and midday heat. As summer progresses, I rise earlier and earlier and settle into my night's rest later and later, unwilling to miss a moment of light or a single portion of summer's riches.

Summer magic. Seductive sun.

The flip side of the North's long dark winters is the glory of its summer sunlight. At Kluane Lake, I climbed down the ladder from our cabin loft in the middle of the night without needing any sort of light, lifted my infant son from his bassinet, and nursed him while seated in an armchair beside a window that looked west toward the Slims River and Sheep Mountain. In dusky light that lingered after the midnight sunset I watched northern harriers glide over a shrubby meadow, their V-wings echoing the sharp angles of the mountains.

Even farther north, at Alexandra Fiord, the sun didn't really set at all during the summers. In mid-August it dipped behind the mountains of Thorvald Peninsula for about ten minutes at 1 a.m., then reappeared and continued circling the horizon. That unending light posed an irresistible temptation to researchers. Spurred on by the compressed High Arctic growing season, we worked long days, often staying up until midnight. June gave way to July, and we stayed up even later, into the wee hours, drugged by the endless golden sunshine bathing the lowland. July gave way to August, and we grew weary, giddy, and sometimes reckless. Once, after working until 2 a.m., we hiked kilometres over the lowland, crossed rivers, and then climbed on a glacier while wearing sneakers, rubber boots, or hip waders—totally inappropriate footwear for a glacier walk. We might have slid into a crevasse but fortunately didn't.

The days wore on, and exhaustion claimed us. The following is an entry I wrote in my journal two weeks before leaving Alexandra Fiord for the last time at the end of August, 1981: *August 15: Time rolls on inevitably, but my energy does not. The weather does not cooperate with us. I am getting burned out. We all are. It's hard for anyone to work after supper now, let alone until after midnight. Perhaps it is good that the end of the season is near—we cannot keep up this pace much longer. Even the plants are burned out from their mad rush of growth while the summer sun was high.*

Summer magic. Seductive sun.

Back in the yard, evening grosbeaks and chickadees call. Overhead, grey cloud rips into yellow-edged tufts that float in a bright blue sea.

March 15 (-3°C, snow flurries, strong NW wind, 4:30 p.m.)

Wind whips at dark green tarpaulins covering our aspen lumber stacked behind the coop. While Dainis and Jānis romp through the yard, Vilis and I reposition the tarpaulins, and Vilis ties them tightly. After half a year of drying, the wood no longer stinks like urine.

"Magi?" Panic laces Dainis's voice as he and Jānis run out from behind the house. "Where are you? *Mag-iii!*"

"Over here," Vilis and I shout, and Vilis adds softy, "I wonder when he'll get over that."

Dainis was three when he disappeared from our yard on a bright winter day six years ago while I dressed his squirming nine-month-old brother for a sled ride.

"I want to go outside *now*," Dainis insisted.

"Okay. Just stay close to the house. I have to finish dressing Jāni, then I'll be right out."

Dainis went out, and I heard his thumps on the old porch wall as I struggled with Jānis's snow pants, and then with boots, mittens, coat, toque, and scarf. I couldn't see Dainis because at that time our kitchen was enclosed on two sides by an old sunporch the previous owner had boarded up completely except for two small windows above child height. Once a week, if not once a day, I told Vilis that when spring came, I was going to rip that porch apart so I could *see.* I set Jānis in our sled and pulled it out of the enclosed porch onto glaring snow, expecting to see Dainis waiting for me. But he wasn't. I called his name and searched all around the house, panic rising within me. Still no Dainis.

Then I heard screaming. The sound was far away and so filled with terror that the hairs on the back of my neck lifted. "Dainis, where are you?" I shouted, but only his faint screams replied. Had he wandered down to the brook and fallen in? Had the coyote we saw on the back road yesterday returned and dragged him away? Fear tore at me. Then I saw him, a small figure in dark bulky clothes, running toward the house from the edge of the woods a quarter of a kilometre away down the back road. I grabbed the sled rope and ran to meet him, bouncing Jānis along behind me. Dainis ran into my arms, tears streaming down his face as he sobbed hysterically, "Mama, where *were* you?"

I held him tight until he stopped crying, then unzipped his coat and snow pants so he could pee in a snowbank. Safe in my arms, he even giggled a bit at the steaming hole and yellow snow. I zipped him up and explained that it had taken a long time to dress Jāni. Had he thought I'd gone down the back road without him? He nodded. And did he get scared when he reached the woods and saw I wasn't there? He nodded again, his face starting to crumple. "Well, let's all go for a walk now," I said, hoping I sounded calmer than I felt.

So down the back road we walked. I pulled the sled and held Dainis's hand as we followed his little footprints in the snow on and on toward the woods, knowing that when he made them, he thought I was somewhere ahead of him. When the dark hemlock woods closed in around us, the footprints stopped, shuffled ahead a bit, and then turned back sharply, the distance between the prints lengthening as my son ran, screaming, trying to escape the realization that I wasn't there, and that he was utterly alone.

"Mama," he asked at bedtime that night, his eyes haunted, his mouth working to put words to his feelings, "was I lonely? When I was in the woods?"

"Not lonely, honey. Alone. And frightened. You didn't know where I was."

Even now, six years later, he still calls my name with panic in his voice. And then we remember.

After Vilis secures the tarpaulins, we all walk the old wood road atop the steep bank beside Matheson Brook, following the path I traced nearly a year ago when I set out on a walk of freedom. We're searching for the bridge but see no sign of it immediately downstream from where it once offered passage across the brook. "Maybe it got caught where the brook makes that sharp turn," Vilis suggests, referring to what I call the Shepherd's Crook.

We hurry to where the wood road makes a sharp turn to the west and look down the bank. And there it is—what's left of the bridge—torn asunder by raging water and piled onto shore at the first sharp turn of the Shepherd's Crook. We scramble and slide our way down the steep slope for a closer look. Some of the deck boards are still nailed to thick support beams; others are torn loose from one beam yet remain attached to another. A huge log with deck boards nailed to its upper end drags its lower end in the water. The rest of the destroyed bridge rests on shore a metre above water level. Above the debris, grasses and water-washed leaves hang from shrub and tree branches like streamers or pew markers flagging the height of the rain-swollen brook.

We scramble back up the slope, clinging to tree trunks, slipping on rocks. Our homeward route takes us past the Amphitheatre and then uphill and through our parkland. At the top of the old wood road, we marvel at hare-clipped apple and pincherry branches, some of which are two metres off the ground. This seems an impossibility until we recall how ice and snow weighted branches until they drooped low, making them easy targets for hares.

Back at the house, we stand by the porch, with clouds a broken shield of grey-blue power above us. Grey hills swell on the southeastern horizon, their edges blurred and mysterious, as though more hills might rise behind them, as though mountains—sharp-shouldered and snow-peaked—might soar above them and lead us forever onward.

March 16 (-8°C, mostly clear, calm, 7 a.m.)

The moon is a pale torn orb in the southwest sky. Brilliant sunshine flares from behind clouds in the southeast. Alders, apples, and spruces at the edge of the yard wear thin stockings of frost. Blue jays squawk, and black-capped chickadees bounce two-note mating songs back and forth against the sky.

It's a different story in the woods. Other than the sound of my footsteps on crusty snow, silence reigns. Slivers of sunlight lie draped across the road like thin orange ribbons. Near the top of the back road, a nuthatch calls its nasal "meep," but its voice sounds weary, only half-awake. The sun hasn't yet reached it, hasn't yet touched the treetops with its light and warmth and awakening.

On Balmoral Road, the world is lit, the sun's golden rays turning the rich greens of spruce and pine to olive. Small streams that feed Matheson Brook are loud and filled with spring importance. They tell of the torrents they became last week when one of them, fed by water rushing down the clear-cut slope above it, poured over the pavement and tore huge bites from the road's shoulder, necessitating the arrival of a backhoe and warning signs for motorists.

From Balmoral Road, I make a brief detour to Balmoral Grist Mill, a working museum perched on the side of a rocky gorge carved by Matheson Brook. At the mill, the brook—black, loud, laced with white-green froth—pounds through the gorge. A twenty-metre spruce tree is jammed under the walkway that crosses the brook above the spillway, no doubt deposited there last week by the turbulent, flooded stream.

The assistant miller once told me that this mill, completed in 1874, could never again be operated with a turbine like it originally was because now the run-off from surrounding logged hills is too fast. A hundred years ago, the mill's turbine worked because the brook's volume was fed by slower, sustained runoff that seeped through forested land and was relatively constant. David Suzuki explained the process this way in *The Sacred Balance*: "That tangle of tree roots snaking across the forest floor absorbs water while holding the soil so effectively that creeks don't flood and the water flowing in them is clean and clear after many days of rain. Held in the soil, in roots and trunks and branches, the water is slowly metered out over days and weeks, and any excess is returned to the air."[46] Thus, our forests are the guardians of our rivers. Here, unfortunately, that guardianship has been destroyed by clear-cutting on the hills above Matheson Brook. I have a small consolation. At Ravenhill we have parkland and meadows transitioning to forest, and that forest will one day exert at least a small calming influence on the boiling brown fury that Matheson Brook becomes after rain.

I complete my swing through the mill grounds and, since our bridge is gone, retrace my steps to the back road. Balmoral Road looks different from this side, in the same way that life looks different from another person's perspective. Water trickles, deep beneath ditch snow, as if it were a genie tinkling out secret laughter. A wind picks up out of the west and pokes cold needles through my knitted sweater, so I pull the anorak I removed at the grist mill back over my head.

On the back road, sunlight now touches the treetops, and the thin orange slivers of light have transformed into broad white bands striping the road. While behind me snow on the pavement lay quiet and dignified underfoot, here in the woods it's crunchy, loud and rambunctious like a boy stomping on puddle ice and thumbing his nose at stuffy adults who watch him and glare. Other than the noise of my passage, silence still reigns, but only until I exit the forest. Then the hushed woods give way to talking meadows where blue jays, chickadees, and evening grosbeaks toss bright bird words against the sky.

5
Letting Go

March 20 (-4°C, overcast with snow, strong NE wind, 6:55 a.m.)

New snow coats our south meadows and parkland, turning the world white and painting hoary moustaches on spruce trees. In that place of change at the edge of the woods, a narrow trail of footprints, perhaps made by an ermine, crosses the back road's smooth throat like a string of beads. In the forest, snow splashes in front of my rubber boots in the same way the Northumberland Strait splashes in front of my bare feet on Rushton's Beach. I'm wading in a snow sea, walking the beach of winter for the last time this season. Spring will officially arrive at 3:55 p.m. today.

Once I walked my usual loop in a winter storm, with snow like this almost mesmerizing me as it sprayed out in fans ahead of my brown leather boots. On that day, the thought came to me that I too often needlessly fill my mind with turbulent thoughts and fret about what I should be doing or what I can't do, and this accomplishes nothing. That day I gazed peacefully at snow falling thickly among the backroad hemlocks and was reminded of Frost's "Stopping by Woods on a Snowy Evening," the poem's words sweeping up from my memory, brought by "easy wind and downy flake."

Initially that storm's snow *was* downy—as light and buoyant as gossamer feathers. Alongside Balmoral Road, a growing wind nudged branches and sent clouds of the fallen feather-snow billowing across the pavement. An SUV passed me from behind, and I didn't hear it until it was close, its engine and tires muted by the snow-filled air and fifteen centimetres of snow on the pavement.

Then the east wind flexed its muscles again, and feathery flakes gave way to shards of icy snow that stung my face. On Matheson Brook Road, I waded through unmarred whiteness until a pickup drove past, the snow spurting from its tires leaving curving twin gouges. I stepped into a track and nearly stumbled, my steps wobbly, my knees—relieved of the burden of pushing snow ahead of my feet—at first lifting too high and free before they adjusted.

Wind shook snow from a neighbour's pines, filling the space around me with white until it seemed that the very air that I breathed was snow. Hardwood trees raised bare limbs to the sky. Pines with flaring, wind-sculpted branches stood stalwart against the storm, like refuges. Aldo Leopold wrote, "The only conclusion I have ever reached is that I love all trees, but I am in love with pines."[47] On that stormy winter day, so was I. Nearly alone on the roads, I gave myself to the snow and wind and pines, content to simply be part of the landscape, a slow-moving figure in a white world.

Today I'm again a slow-moving figure in a white world, but this world is no longer completely the domain of winter. Already I've experienced the vernal sun's enticing power as the rate of day length increase approaches its maximum leap on the sine wave curve, on the upward-flicked ribbon. Yesterday, as I was returning from a walk, sunshine bathed these woods, touched my face, and halted my steps with its caress. I sat on a hemlock stump and gave myself to the radiant warmth, lifting my face to its golden kiss. All else was forgotten—ice, snow, work, family. I felt only the sun's comfort and heard only the whisper of spring.

March 22 (-4°C, overcast with ice pellets, strong NE wind, 7:55 a.m.)

Ice pellets sting my face, but in the excitement of this day, they are nothing. I've tried on another new coat and like the feel of it. Last month, before I first sat down in a CBC Radio studio in Halifax, I was nervous. After the recording session, I trembled. Yet all else faded while I read my story about how a flower box filled with pansies on a Yukon cabin helped me transition from biologist to writer.[48] During that recording, nothing existed except the words and me. My voice in the earphones startled me, for although I knew it was my mouth speaking, I sounded like someone else—an exotic stranger. Instantly, I fell in love with radio.

Now, after that first dance, CBC has held out its hand and asked me for another, in the form of a collage of vignettes about Dainis and Jānis, which I call "Out of the Mouths of

Boys."[49] So, this morning I walk with the new coat wrapped close around me, and the ice pellets stinging my face are nothing.

March 26 (-3°C, clear, W breeze, 7:30 a.m.)

In early morning, I stride over wind-packed snow in the parkland north of our yard. Beside me, Dainis and Jānis bubble with excitement. They're excited for two reasons. First, we're on our way to collect maple sap. Second, this is the first time they've ridden their bicycles on snow. They're astonished by the wind-packed surface's firm traction and thrilled with the smooth expanses the snow has created in otherwise bumpy parkland. They pedal on ahead of me, circle back, and explore clearings through which I pulled our toboggan yesterday en route to the Amphitheatre. Vilis is away at a conference, so it falls to me to bring the sap home; yet, it was Dainis and Jānis who urged me out of bed yesterday and this morning and hurried me through breakfast so we could check the buckets.

In the forest, snow is too soft and loose for cycling, so the boys leave their bikes at the top of the old wood road. When we reach the Amphitheatre, they run down the steep bank to see how much sap has dripped into their buckets. They each have two within the Amphitheatre, and Vilis and I have five buckets on trees growing along the top of the bank. These nine taps are a pittance compared with any serious maple producer, but they yield enough syrup to see us through most of the year's pancake and waffle breakfasts.

Today, some buckets are a third full, some are four-fifths full, and some are almost empty. I pour the sap into the turquoise container, pausing to scoop out flattened caps of ice, one of which I give to Jānis so he can suck maple essence from its crystals. Then I lash the plastic container and our coats to the toboggan, and together we tackle the Amphitheatre's slippery wall. I pull the load, Dainis pushes and braces it, and Jānis wraps a long lead rope around a slim tree upslope to prevent our precious cargo from escaping. A little sap leaks onto Dainis's clothes, and we have to rest partway up the slope with the toboggan braced against the base of a beech tree, but then we push on. Excelsior!

When we reach the top, we're all panting, but we're laughing, too. Here we are, free people out on the land, with sunshine brilliant on snow and the taste of maple sap on our lips. The boys romp and natter beside me as I haul the toboggan up the old wood road. At the top, they lift their bikes, walk them to hard-packed snow, and ride.

March 27 (3°C, overcast, W wind, 6:30 a.m.)

A new voice rings out in the yard—a chipping sparrow rattling out its metallic song. Juncos trill, and robins stride over our lawn with heads high, looking and listening.

I stride, too, and run, easing feet and ankles, knees, hips, and back into the progression from firm walk to slow, careful run. Even this light jog offers instant freedom, instant power over the earth beneath my feet. I eat up the back road and Balmoral Road, my feet grabbing gravel, my lungs sucking in and pumping out air. Run, walk; run, walk. A sharp wind bites my face, cold embraces me, and my sneakers crunch brown leaves caught in roadside stubble. I'm not caught in roadside stubble. I'm running.

March 30 (1°C, mostly clear, calm, 6:40 a.m.)

Juncos trill from every direction, and a robin sings its cheerful song. In the yard, Vilis lights a fire under a pot of maple sap while I head down the back road to run and walk intervals, granting myself forty minutes of aloneness, movement, and awakening.

Thus, the rituals of spring begin. Rendering maple sap into syrup. Running. Drying laundry outdoors. Raking gravel off the lawn. Collecting and burning bark pieces that flew from the splitting maul. Cutting a vase of pussy willows. Gathering garbage blown about by winter winds. Cycling on soft gravel. Relaxing on sun-warmed porch. Releasing chickens into the yard. Dashing to see if crocuses are open. Letting go of winter.

I run and walk the back road and Balmoral Road as far as Matheson Brook Road before retracing my steps. For the first time this year I hear winter wrens' bright bubbling songs and a flicker's sharp "kek-kek-kek" in the backroad woods. In the Cathedral's shadowed grace, my back straightens and my soul quiets. Past the Cathedral, I marvel at how much ice has melted since I slipped and skidded my way along this road yesterday. A red squirrel chatters, and I hear a nuthatch's nasal call. Splinters and panes of ice coat puddles in that place of change at the edge of the woods. Then I'm in the meadows again, with the face of our old house—weathered and warmed by sunshine—beckoning me home.

As I step onto the porch, the patio doors reflect the figure of a woman and the scene behind her—large flowerbed, bird feeder on a pole, lawn and apple trees, young spruces in a meadow, hills rising in smooth curves on the horizon. The woman wears running shoes, and sunshine on her hair spins it into a golden halo. In this instant, it's as though I'm seeing her for the first time, as though I'm seeing *me* for the first time. And yet I know without a shadow of a doubt that the blonde woman reflected in the glass—a woman who walks and seeks, who runs and listens, who dreams and writes—is truly who I am. She is the one I was seeking a year ago when I set out on an old wood road atop a steep slope beside Matheson Brook. And now she's here. Now *I'm* here.

Who can say where we find ourselves? I found myself in air and soil and water, in frost thorns and in red maple leaves that fell onto wet, dying green. I found myself in my sons' laughter and in the touch of my husband's shoulder beneath my hand as we danced. Throughout all, the earth nourished me, freed me from myself, then gave me back myself. My "faint tracing on the surface of mystery" is a little bolder, a little brighter.

April 4 (2°C, overcast with snow, strong N wind, 3:45 p.m.)

In late afternoon, I abandon the house and my writing. Wind shoves at my back, and snow dusts hemlocks beside the back road. Branches call to my eyes, telling stories I don't see in summer. Canid tracks—fox, I believe—print neat round designs in thin snow. I notice drag marks on the fox's left, sign that it was carrying prey whose fur or feathers slid over the snow. I lose the trail on bare ground under a hemlock, my eyes not skilled enough to follow it into the forest.

Today is a day of light and dark, snow and bark, and green moss lush beneath cloudy skies. Snow pelts my face as I reach Balmoral Road. I turn right and walk on into the rest of my life.

Dear Reader,

Thank you for reading *Seasons on Matheson Brook: A Journey to Self Through Exploring Nature*. I hope you enjoyed it. Please consider leaving a star rating and honest review (as short or long as you like) at Goodreads or the retailer where you purchased this book. Reviews are vital for helping readers discover good books in their area of interest. I would very much appreciate your taking the time to post a review of *Seasons on Matheson Brook*.

Thank you!

Magi Nams

Author's Endnote

My love affair with the natural world and desire to share it with others didn't end on April 4, 1998. I continued to walk and run and explore, and continued to write articles for *Ranger Rick*, as I squeezed in slivers of freedom and discovery among my responsibilities.

Two years after my journey to self on the banks of Matheson Brook, my family had the opportunity to spend ten months in New Zealand. That time of exploring new landscapes and ecosystems was a dream come true, a once-in-a-lifetime adventure. We travelled throughout much of the country, hiked hundreds of kilometres, did scientific research, and immersed ourselves in Kiwi life.

Again, I was driven to write, with words arising from the heart like a love song. Eventually, I wove my "NZ stories" into a travel adventure series, *Cry of the Kiwi: A Family's New Zealand Adventure*, which chronicles the joys and challenges, struggles and awakenings experienced by a Canadian family in a land of volcanoes, giant trees, mammalian pests, and endangered flightless birds. For your enjoyment, I've included the opening of the first book in the series, *Once a Land of Birds*. Cast your gaze on the next page and meet New Zealand.

Cry of the Kiwi: A Family's New Zealand Adventure Book 1, *Once a Land of Birds*

August 9, 2000

Orange streaks flare against the night like a beckoning light at the end of the world. I stare through the window at waning darkness while my husband, Vilis, and sons, Dainis and Jānis, sleep onboard the Air New Zealand jet. A strip of cloud below the jet looks like thick smoke, tinged with the colour of ripe muskmelon. Through gaps in the cloud, I see steel-blue waves like beaten metal, and brown peaks like the humps of stampeding bison. A thrill runs through me. *This is it!* These are my first glimpses of Nieuw Zeeland (so named by Dutch geographers for their coastal province of Zeeland). To the Māori, whose ancestors arrived from Polynesia centuries before the Dutch explorer Abel Tasman sailed in search of a southern continent in 1642, this is Aotearoa, "Land of the Long White Cloud."

At 5:00 a.m., a lingering darkness cloaks Auckland International Airport. Groggy from jetlag after flying half way around the world, my family trudges through New Zealand Customs. We're detained by a sniffer dog, a beagle who's detected the scent of fruit emanating from my daypack. If I had any fruit, it would have been a contravention of New Zealand's strict biosecurity regulations intended to prevent pests and diseases from entering the country, however I ate the last dried apricot somewhere over the Pacific Ocean. We're released from the dog's inspection and follow directions that lead us through the airport and beyond its doors. The caress of warm air in the southern winter startles us.

"Look! There's a *palm* tree!" Jānis's voice is filled with wonder.

Despite our weariness, the sight of those gracefully curved, exotic leaves sparks excitement in us. We're four Canadians 18 000 kilometres away from home (an old farm in Nova Scotia) where palm trees definitely do not grow outdoors!

This is why we came. This is why we left the familiarity of Nova Scotia for the newness of New Zealand. We had a need to explore a part of the world far away, to see new plants and animals and landscapes, to seek new adventures and feel new experiences jolt our senses and expand our horizons. For years, as a homeschooling parent, I'd wanted to live with my husband and kids in a foreign country. Such an adventure would be the ultimate fieldtrip. As the only ecologist on staff at the Nova Scotia Agricultural College, Vilis had for years looked forward to taking a leave to focus exclusively on his research, with no teaching responsibilities. When the New Zealand possibility arose, he and I leapt at the chance. And our sons? Jānis radiated excitement at the prospect of travel adventures. Dainis was torn between the contentment of remaining in Nova Scotia with his beloved home, friends, and Lego, and the desire for new experiences.

Those new experiences start now, and it's Vilis's research that's made them possible. We've come to New Zealand because of an animal Kiwi conservationists wish didn't exist in their country: the stoat. A stoat is a kind of weasel. It's been declared public enemy number one of New Zealand's native birds, particularly this country's national icon and unofficial symbol, the kiwi. You know the kiwi: dumpy and flightless, almost blind, and with a long beak like a poker. It's the reason New Zealanders are colloquially called "Kiwis." All five kiwi species face extinction, and stoats, which prey on kiwi chicks, are one of the main threats to kiwi survival. This is where Vilis, who has experience live-trapping and radio-tracking weasels and other predatory mammals in Canada, enters the picture. He'll assume a short-term position with Landcare Research, a government agency that studies and manages terrestrial ecosystems. His research will be aimed at helping to rid New Zealand of stoats.

As we stand in the warm Auckland darkness staring at a palm tree, I feel my family's horizons stretch. What adventures will the next ten-and-a-half months bring? For one of those months, Dainis, Jānis, and I will help Vilis live-trap and radio-track stoats in the Tongariro Forest Conservation Area on North Island, the northernmost half of this country. During the remainder of our sojourn, we'll be based near the Landcare Research branch in Lincoln, a township twenty kilometres southwest of Christchurch in the South Island province of Canterbury. There, as a family, we'll embark on a homeschooling and outdoor adventure odyssey intended to immerse us in New Zealand's history, culture, landscapes, flora, and fauna. As for me personally, I have three goals for my time in New Zealand: lots of hiking, lots of birding, and lots of writing. We've been fortunate to receive the travel and educational opportunity of a lifetime, and I plan to make the most of every minute…

List of Illustrations

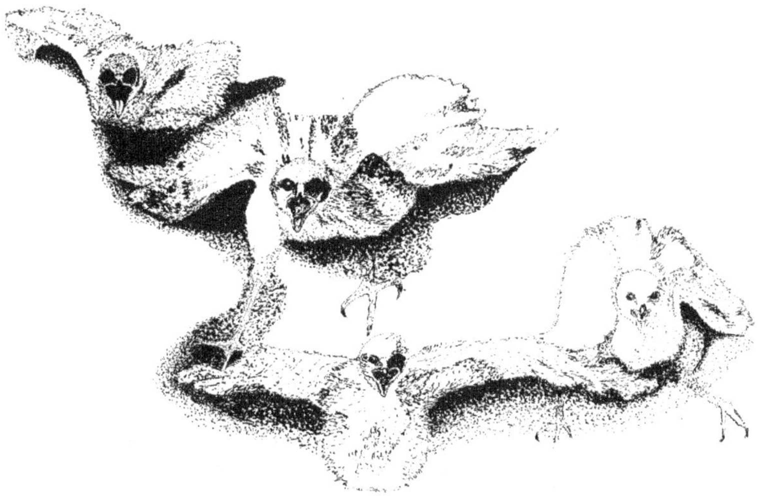

Acknowledgements

It's been decades since I penned the first words of my journal, and without the assistance and encouragement of family and friends, this at-last-completed book would not be in your hands. First and foremost, loving thanks to my husband, Vilis Nams, whose offer to share teaching responsibilities granted me freedom to explore the world on my doorstep and time to write about how my observations impacted my life. Insightful comments by Vilis and fellow beta reader, Anne Mains, allowed me to take an early draft of this book to the next level. Special thanks to Nancy Drysdale and Glen Matheson for providing me with the genealogy of the Hayman family who settled what is now my family's property. I also had the good fortune to have Brian Bartlett copyedit my manuscript and offer helpful suggestions as to how to trim, clarify, and vivify it. Last, but never least, I thank Vilis and our sons, Dainis and Jānis, for allowing me to share their roles in my personal journey with the world. They are the men of my heart, and this book is for them.

Other Books

by Magi Nams

Cry of the Kiwi: A Family's New Zealand Adventure series:

Once a Land of Birds

This Dark Sheltering Forest

Tang of the Tasman Sea

written under pen name Katie Ardea

Prairie Hearts series:

Braver Than You Know

References

[1] John Muir. *Travels in Alaska.* 1915. Houghton Mifflin Company, Boston, p. 5.

[2] Barbara Kingsolver. *The Poisonwood Bible.* 1998. Harper Collins Publishers, Inc., New York, p. 5.

[3] Eden Phillpotts. *A Shadow Passes.* 1918. Cecil Palmer & Hayward, London, p. 19.

[4] Lewis Carroll. *The Annotated Alice: Alice's Adventures in Wonderland & Through the Looking Glass.* 1960. NAL Penguin, Inc., New York, p. 179.

[5] Thomas Morell (translator). 'There's a certain Divinity in good Men. A man is not to be esteemed for any external and foreign Good.' Epistle XLI in *The Epistles of Lucius Annaeus Seneca*, Volume 1. 1786. W. Woodfall, London, p. 142.

[6] Eugene H. Peterson. *The Message: The Bible in Contemporary Language.* 2002. NavPress Publishing Group, Colorado Springs, CO. Romans 1:20, p.1305.

[7]NASA Astronomy Picture of the Day. https://apod.nasa.gov/apod/ap050522.html.

[8] Aldo Leopold. *A Sand County Almanac: with other essays on conservation from Round River.* 1966. Oxford University Press, New York, p. ix.

[9] Edward Abbey. *Desert Solitaire: A Season in the Wilderness.* 1968. Simon & Schuster, New York, p. 169.

[10] Annie Dillard. *Pilgrim at Tinker Creek.* 1974. Harper & Row, Publishers, Inc., New York, p. 143.

[11] Marshall Brickman and Rice Elice. *The Addams Family: A New Musical Comedy*, Cast Script & Vocal Book (Music and Lyrics by Andrew Lippa). 2009.Theatrical Rights Worldwide, New York, p. 24.

[12] Biblica, Inc. *Holy Bible, New International Version.* 2011. Zondervan, Grand Rapids, MI, Genesis 3:19, p. 4.

[13] Mary Oliver. "The Summer Day" in: *New and Selected Poems, Volume One.* 1992. Beacon Press, Boston, p.94.

[14] Paul R. Ehrlich, David S. Dobkin, and Darryl Wheye. "MacArthur's Warblers," in *The Birder's Handbook: A Field Guide to the Natural History of North American Birds.* 1988. Simon & Schuster Inc., New York, pp. 523, 525.

[15] Hance, Roy Ivor. "The Enigma of Bird Anting," *The National Geographic Magazine*, July 1956, pp. 105-118; also "Anting"' in Ehrlich, Dobkin, and Wheye, p.487.

[16] Maureen Houston. "Grouse and predators researched in Turtle Mountains," *The Recorder*, Boissevain, Manitoba, June 20, 1984, pp. 8-9.

[17] Tinbergen L. "The natural control of insects in pine woods I. Factors influencing the intensity of predation by songbirds." Archives Néerlandaises de Zoologie. 1960. 13:265–343. Cited in "Visual Search and Attention in Blue Jays (*Cyanocitta cristata*): Associative Cuing and Sequential Priming." Kazuhiro Goto, Alan B. Bond, Marianna Burks, and Alan C. Kamil. *Journal of Experimental Psychology: Animal Learning and Cognition.*, 2014. Apr. 40(2):185-194.

[18] Rachel Carson. *The Sense of Wonder.* 1965. Harper & Row, Publishers, New York, p. 45.

[19] Erich Fromm. *The Anatomy of Human Destructiveness.* 1973. Holt, Rhinehart and Winston, New York, p. 406.

[20] Edward O. Wilson. *Biophilia: The Human Bond with Other Species.* 1984. Harvard University Press, Cambridge, MA, p. 1.

[21] B. Freedman, J. Svoboda, C. Labine, M. Muc, G. Henry, M. Nams, J. Stewart, and E. Woodley. "Physical and ecological characteristics of Alexandra Fiord, a High Arctic oasis on Ellesmere Island, Canada." 1983. Permafrost, Fourth International Conference Proceedings, July 17-22, 1983, pp. 301-304.

[22] Biblica, Inc. *Holy Bible, New International Version.* 2011. Zondervan, Grand Rapids, MI, Psalm 24:1, p. 713.

[23] Madeleine L'Engle. *A Circle of Quiet.* 1972. Harper & Row, San Francisco, p. 209.

[24] Henry David Thoreau. *Quoted in*: Robert Epstein and Sherry Phillips. *The Natural Man: Henry David Thoreau.* 1978. Quest Books, Wheaton, IL, p. 75.

[25] Paul R. Ehrlich, David S. Dobkin, and Darryl Wheye. "Hovering Flight," in *The Birder's Handbook: A Field Guide to the Natural History of North American Birds.* 1988. Simon & Schuster Inc., New York, pp. 323, 325.

[26] Paul R. Ehrlich, David S. Dobkin, and Darryl Wheye. "Ruby-throated Hummingbird," in *The Birder's Handbook: A Field Guide to the Natural History of North American Birds.* 1988. Simon & Schuster Inc., New York, p. 328.

[27] Annie Dillard. *Pilgrim at Tinker Creek.* 1974. Harper & Row, Publishers, Inc., New York, p. 12.

[28] R. S. Wilcox. "Communication by surface waves: Mating behavior of a water strider (Gerridae)." 1972. *Journal of Comparative Physiology* 80: 255-266.

[29] R. S. Wilcox. "Sex discrimination in *Gerris remigis*: role of a surface wave signal." 1979. *Science* 206: 1325-1327.

[30] Albert Einstein. *Quoted in*: William Hermanns. *Einstein and the Poet: In Search of the Cosmic Man.* 1983. Branden Press, Brookline Village, MA, p. 138.

[31] University of Wisconsin-Madison, Wisconsin Horticulture, Division of Extension. Goldenrod Gall Fly, *Eurosta solidagnis.*

https://hort.extension.wisc.edu/articles/goldenrod-gall-fly-eurosta-solidagnis/.

[32] Aldo Leopold. *A Sand County Almanac: with other essays on conservation from Round River.* 1966. Oxford University Press, New York, p. 81.

[33] Bernard Campbell. *Human Ecology: The Story of Our Place in Nature from Prehistory to the Present*, second edition. 1995. Aldine de Gruyter, New York, p. 209.

[34] Smithsonian Institution. "Numbers of Insects (Species and Individuals)." BugInfo, Information Sheet Number 18. Prepared by the Department of Systematic Biology, Entomology Section, National Museum of Natural History, in cooperation with Public Inquiry Services, Smithsonian Institution. https://www.si.edu/spotlight/buginfo/bugnos.

[35] Biblica, Inc. *Holy Bible, New International Version.* 2011. Zondervan, Grand Rapids, MI, Psalm 96: 11-12, p. 775.

[36] Jonas Salk. Address on receiving the Nehru Award (10 January 1977), published in *Virginia Woolf Quarterly* (1977), Vol. 3, p. 11.

[37] Robert Bateman. *The Art of Robert Bateman.* 1981. Madison Press Books, Toronto, p. 59.

[38] Robert W. Pennak. *Fresh-Water Invertebrates of the United States.* 1953. The Ronald Press Company, New York, p. 506.

[39] *Ibid.* p. 515.

[40] Mother Teresa. *In the Heart of the World: Thoughts, Stories & Prayers.* 1997. New World Library, Novato, CA, pp. 20-21.

[41] Charles J. Krebs, Rudy Boonstra, Stan Boutin, and A.R.E. Sinclair. "What Drives the 10-year Cycle of Snowshoe Hares?" 2001. *BioScience*, Volume 51, Issue 1, 1 January 2001, Pages 25–35.

[42] *Ibid.*

[43] *Ibid.*

[44] William Blake. "Auguries of Innocence." https://www.poetryfoundation.org/poems/43650/auguries-of-innocence.

[45] Robert Frost. "Birches," in *Robert Frost: Selected Poems*, 1975 edition. Penguin Books Ltd., Hammondsworth, p. 80.

[46] David Suzuki. *The Sacred Balance.* 1997. Douglas & McIntyre, Vancouver, p. 55.

[47] Aldo Leopold. *A Sand County Almanac: with other essays on conservation from Round River.* 1966. Oxford University Press, New York, p. 74.

[48] Magi Nams. "The Flower Box." CBC Radio, *First Person Singular*, March 5, 1998.

[49] Magi Nams. "Out of the Mouths of Boys." CBC Radio, *Out Front*, May 27, 1998.

Made in the USA
Monee, IL
04 November 2026

67386737R00154